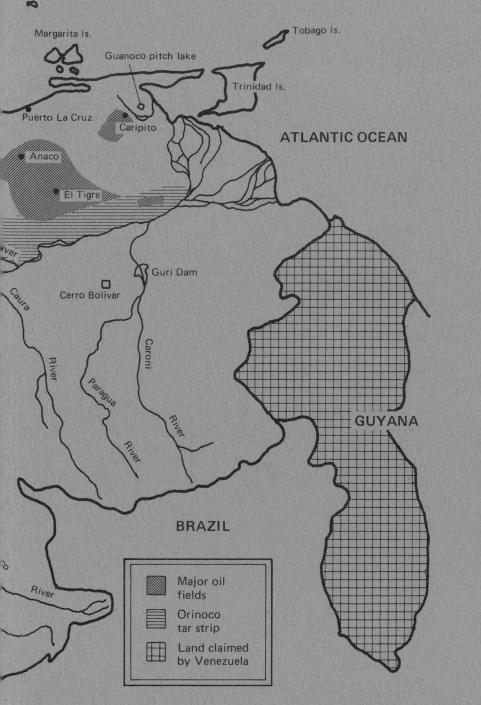

Grenada Is.

S0-AXK-769

Margarita Is.

Tobago Is.

Guanoco pitch lake

Trinidad Is.

ATLANTIC OCEAN

Puerto La Cruz

Caripito

Anaco

El Tigre

River

Guri Dam

Caura

Cerro Bolivar

Caroni

River

Paragua

River

River

GUYANA

BRAZIL

River

Major oil
fields

Orinoco
tar strip

Land claimed
by Venezuela

Oil: The Making of
a New Economic Order

Oil: The Making of a New Economic Order

Venezuelan Oil and OPEC

LUIS VALLENILLA

McGRAW-HILL BOOK COMPANY

New York   St. Louis   San Francisco   Auckland   Düsseldorf
Johannesburg   Kuala Lumpur   London   Mexico   Montreal
New Delhi   Panama   Paris   São Paulo   Singapore
Sydney   Tokyo   Toronto

Library of Congress Cataloging in Publication Data

Vallenilla, Luis, date.
    Oil, the making of a new economic order.

    1. Petroleum industry and trade—Venezuela.
2. Organization of Petroleum Exporting Countries.
I. Title.
HD9574.V42V36   338.2'7'2820987   75-22447
ISBN 0-07-066830-2

1234567890   BPBP   7321098765

This book was set in Baskerville by University Graphics, Inc.

Printed and bound by The Book Press.

To Anita

Contents

Preface

Developed nations are proclaiming the existence of a "petroleum crisis" that threatens to strangle their economies because of the substantial increase in oil prices established by petroleum-exporting countries toward the end of 1973. The public is being led to believe that this higher cost of petroleum is the main factor causing the inflation unleashed in their countries and draining their financial resources and that it results from the prodigious flow of funds to oil-exporting countries.

From another viewpoint it may be said that the crisis surrounding petroleum today is not a product of the current situation but, rather, that it has always existed, only the parties involved have become inverted. This view asserts that it was the developed nations that retarded the development of the economies of petroleum-exporting countries, by paying low prices for this nonrenewable resource and at the same time increasing—year after year—the prices of their own finished export products.

Formerly it was the petroleum countries that frequently had deficits in their trade balances with the United States, Europe, and Japan, while having to watch the ever-increasing and unceasing shrinkage of their precious energy resources and confronting the most varied economic and social problems. Nothing was said by the industrial nations at that time of the existence of a petroleum crisis or strangulation of economies, possibly because their communities were not the parties affected by the prevailing trade imbalance.

In any case, both points of view are a simplistic approach to problems that should be profoundly analyzed, bearing in mind all contributing factors.

Petroleum countries have traveled a long way toward attaining improved prices for their hydrocarbons—a struggle that is clearly evident in the oil history of any great petroleum-exporting nation. Venezuela, having a petroleum history of slightly over half a century, was in the vanguard as the foremost export country until the international oil companies—backed by industrial nations—intensified the diversification of their sources of supply and developed the industry in the Middle East, which possesses tremendous reserves, and in Africa.

Readers of this book will find certain fundamental opinions expressed from the standpoint of petroleum-exporting countries. The essential aspirations of the Third World in general and petroleum countries in particular are described. My main purpose, however, is to try to enlighten my readers about the oil industry in Venezuela: what it has been, what it is, and what its future might be. Also included are basic aspects of the petroleum-exporting industry in general in view of its international nature.

As mentioned initially, world problems cannot be analyzed simplistically, and therefore it is essential for all of us to understand that present world economic conditions are not fundamentally a consequence of the current price of oil. These conditions have far more complex and profound roots, with causes shaped by many socioeconomic factors on an international scale.

One such critical factor is, in my opinion, the inequitable equilibrium that characterizes international trade. There is a very high degree of interdependency in the world today, and trade relations cannot possibly continue developing on the basis of a small group of increasingly richer and dominant countries on the one hand and a large group of dominated, exploited, and impoverished countries on the other. Rather, these relations should develop on the basis of mutual interest in international justice.

After World War II the economic expansion of the United States and the resurgence of Western industrial powers further delineated the duality of rich and poor, industrial and rural, and developed and underdeveloped nations. The oil structure favoring industrial nations began to weaken at the end of the 1960s, and the concentration of forces shifted in favor of the few oil-exporting countries. The dominant countries of yesterday now harbor fears of becoming dependent, at least in respect to raw materials, on Third World countries. In addition to considering this world problem of today, I have outlined goals and strategies that might be pursued and applied by petroleum countries and supported by the Third World countries in order to rationally consolidate this shift in forces and result in a better economic equilibrium.

If underdeveloped countries are able to untie the bonds of economic dependency to which they are subjected, then, at least theoretically, the world might awaken to a luminous dawn where a more equitable distribution of world wealth prevails and a genuine economic interdependency among all nations replaces the traditional imbalance of power, income, and opportunities, which is a deeply rooted cause of present discontent and grave international tensions.

When the moment arrives in which leaders of all nations—both developed and underdeveloped—comprehend and deeply feel the need to pacifically, rapidly, and constructively overcome the tremendous differences separating them today, we ought to be able to bequeath to our future generations a better, more stable, and peaceful world.

Luis Vallenilla

Acknowledgments

I would like to express my gratitude to Pedro Grases, my good friend and teacher from childhood. He not only contributed to orienting my studies and formation but knew how to persuade me to write this book, with that enthusiasm and vigor which he always imparts to everything that represents intellectual work and an objective contribution to Venezuela.

To my close friend and colleague Dr. Ramón Raven, I also wish to express my gratitude and appreciation for his excellent collaboration, which has been fundamental in the preparation of this work. Without his capacity for analysis, profoundness of concepts, tenacity of investigation, and inexhaustible efforts it would have been difficult for me to complete the present book in its full breadth and dimension.

Rafael Hamilton, my efficient and close collaborator of many years in petroleum matters and Executive Director of our Technical Office Venezuelan Petroleum, has also rendered fundamental aid in the elaboration of this book. The clarity of his petroleum knowledge, his ease with figures, and his dedication, discipline, and sense of organization have contributed in great measure to the accomplishment of this work.

To my father, Pedro Vallenilla Echeverría—to whom, together with my mother, María Celeste de Vallenilla, I owe almost all that I am—go my gratitude for the stimulus that, along with interest, he gave me in the fulfillment of my task. Apart from his love of study, democracy, and free expression of ideas, which he knew how to instill in me from my formative years and which, no doubt, have had great influence in this work, his counsels have been very useful in its preparation.

I also wish to express my profound gratitude to my wife, Anita. Her just evaluation of the scope of this work from its beginning several years ago, her continued interest in it, her infinite patience, and, above all, her affection and comprehension have been the moral support I needed to initiate and pursue the arduous task.

To Professor Manuel Pérez Vila not only am I indebted for an excellent general revision of the book but also for enthusiastic support and wise advice.

Dr. Agueda Agreda deserves my thanks for her valuable aid and interest, as does my secretary, María de Doval, who, in addition to her multiple daily tasks, found the time to type uncountable drafts and collaborate effectively with me.

A word of recognition also to Rocío Márquez for her efficient aid, as well as to my helpers with the English text, especially Elizabeth Tylor. Finally, my thanks to many friends and volunteers, too numerous to mention, who generously gave me their support and valuable advice and shared their knowledge and ideas.

PART ONE

The Past— Development

> To be born and become strong comes first.
> During infancy we need support and in
> manhood we shall have learned how to
> defend ourselves.
>
> SIMÓN BOLÍVAR
> Letter to José Rafael Revenga,
> Magdalena, February 17, 1826.

ONE

Promotion of the Industry (up to 1917)

History

Thirty-five centuries before the birth of Christ, according to evidence found at archeological excavations, oil was known in Mesopotamia—today Iraq. This oil was used as an agglutinant for sealing bricks together and as an element to protect wood from decay. It also seems to have been used frequently in ancient warfare: the Byzantines threw "stones of fire" at enemy ships approaching their shores, and the Assyrians flung boiling oil and "torches" at invaders.

By the Middle Ages, oil was being used in lamps; it was being used as a lubricant and even as a medicine; furthermore, it was the "pitch" widely used for repairing and waterproofing ships. In the history of China, Greece, Rome, Egypt, Burma, Romania—and even in the history of the indigenous populations of America—there exists evidence of its use for similar purposes in remote eras. However, it was not until 1859 that the world's vast oil industry saw the light of day, when Col. Edwin Drake drilled the first oil well in Pennsylvania.

Toward the end of the nineteenth and beginning of the twentieth century, two powerful corporations and two great men triggered the industrial boom and monopolized the oil industry: first came the Standard Oil Company, headed by John D. Rockefeller, and later came the Royal Dutch Petroleum Company, headed by Sir Henri Deterding. In

3

the earliest stages of the industry, an important role was also played by the Nobel Brothers Naphtha Company of Russia—the nation which at the turn of the century produced more oil than the United States. For many years Nobel was a menace to Standard Oil, until the Bolshevist Revolution caused the decline of the powerful Russian oil industry.

In 1870, John D. Rockefeller—a man of outstanding drive—founded the Standard Oil Company of Ohio with a capital of $1 million. Under Rockefeller's administration and thrust, Standard became one of the great international companies contributing to the creation of the vast oil industry, from which its immense fortune was mainly obtained. The company's target was to gain control of oil refining, distribution, and transportation. Standard's major profits were derived from these services.

Rockefeller's shrewdness led his company to a large-scale financial status and a high level of efficiency. For many years Standard was the world's undisputed oil supplier. Its foreign markets were larger than its domestic market, and even in those days Standard's oil was being shipped to the farthest corners of the world.

Standard Oil Company's monopoly prompted the Sherman Antitrust Act, passed by the U.S. Congress in 1890, whereby any contracts or combinations of a monopolistic nature became illegal. Undaunted, Rockefeller reorganized the structure of his company and managed to maintain Standard's primacy in the United States, and also in a large part of the world, for many years. In 1911, Standard of Ohio faced new accusations, and was legally forced to dissolve, the outcome being a new holding company, Standard Oil of New Jersey.[1] The new company controlled all the companies into which the Ohio company had been split, with Rockefeller at the helm.

The boom triggered by Standard aroused in its early years Russia's interest in exploiting her immense Apsheron reserves in the Caspian Sea. Unlike the United States, Russia paid more attention to fuel oil, which it used for heating, transportation, and industrial purposes. In 1873, modern drilling techniques were adopted, thus originating the wealthy Baku gush. Baku acquired major importance as a refining center, to such an extent that in 1884 Russia flourished as an oil-exporting nation, rivaling the United States.

In 1878 the Nobel brothers, Ludwig, Robert, and Alfred, of Swedish origin, founded the Nobel Brothers Naphtha Company. They built the first pipeline (in the Apsheron fields), thus introducing the technique into Russia's newly born industry. They also built the first major tanker fleet. Twenty years were to pass before Standard first employed the continuous distillation already used by Nobel in 1883.

---

[1] In October 1972, Standard Oil of New Jersey changed its name to Exxon Corporation.

Stirred by the Russian menace—and distrusting its own independent distributors—Standard in 1888 created Anglo-American Oil, which later became the head office for all Standard's foreign companies.

Several years before, the oil world had been discussing the feasibility of dividing up world markets among the companies. In 1888, Baron Alphonse de Rothschild (the great Parisian financier who granted credit for building the Caucasus railway in exchange for the right to manage Russia's export surplus) talked with John Archbold, president of Standard, leading to certain agreements between Standard and Nobel Oil involving market sharing and quota assignments. About that same time, some price warfare erupted when Standard suspected the Russians of taking more than their assigned quota.

On another front, Romanian oil had been vied for since 1864, because its quality made it an important gasoline source. Successive governments leased the royal lands to various private interests, and fears arose about the financial independence of the kingdom. A liberal Romanian minister (Sturdza) then proposed that the oil be exploited by a national corporation; this was the first attempt to "nationalize" the industry.

Toward the end of the nineteenth century, Standard acquired Imperial Oil (of Canada), thus capturing 60 percent of the Canadian market; in Mexico it took over the Waters Pierce Company.

At the beginning of the twentieth century, when Standard had apparently survived the Russian menace, the Royal Dutch Company for the Exploitation of Oil Wells in the Dutch Indies, founded in 1890 and later to become the Royal Dutch Petroleum Company, flourished as a giant corporation. Notwithstanding competition from Standard and other difficulties, this company had already reached a high level of development. Thus, Dutch East Indian oil started flooding the Far East. Sir Henri W. A. Deterding was soon to become the prominent figure in this company.

The prosperity achieved by Royal Dutch—especially as a producer—awakened the covetousness of Standard and of Shell (owned by Marcus Samuel,[2] an oil shipper like Standard). However, at first the Dutch nationalistically rejected all kinds of alliances, and to prevent Standard from becoming a shareholder, they declared that only Dutch nationals could buy their stock.

Deterding soon realized that he would eventually have to join Shell to be able to compete with Standard, but his differences with Samuel prevented the deal. Nevertheless, both European companies founded the

[2]Marcus Samuel was the founder of a store in London, Shell Transport and Trading Company, which bought and sold goods—especially seashells—from the Far East. Upon succeeding him as head of the enterprise, his son became engaged in general import-export, including the transportation of kerosene.

Asiatic Petroleum Company, which was to handle their business in the East. Baron Rothschild also participated in this company, and finally two giant companies—Royal Dutch and Shell—merged in 1907.

At the end of the nineteenth century, world production was approximately 120 million barrels per year. Russia's production exceeded that of the United States. Exploitation in the Dutch Indies and in Mexico started up, and a few years later Argentina, Egypt, Venezuela, Romania, and Persia emerged as leading oil sources. In those days oil companies did not have to exert themselves to sell their product because the industrial revolution had already set in motion numerous oil-operated machines—locomotives, ships, automobiles, and airplanes. Thus, the need for oil at all levels of life was to increase endlessly. Oil had already started to displace all other sources of energy, a process which was accentuated with the passing years.

Social disturbances culminating in the Bolshevist Revolution of 1917 marked the decline of Russian oil. "Production began to waver in the Baku fields, the largest the world had ever known, where the Tagiev source produced more oil in a single day in 1886 than the rest of the wells throughout the world, including the 25,000 barrels in the United States."[3]

With the Russian decline, Standard gained strength, but rivalry with the Royal Dutch Company continued. Deterding was always in favor of a price agreement, but Standard used prices for controlling its own positions; this it could do because its grip on the world market enabled it to counterbalance its losses. Furthermore, Standard had an exclusive monopoly of the United States market.

For special reasons Deterding and Rockefeller manipulated their businesses from different points of view. Deterding was always supported by the British and Dutch governments, whereas Rockefeller had rather to defend himself against public and government onslaughts on his monopolistic organization. The United States governmental policy was justifiable because other producers were negatively affected by Standard's power. On the other hand, even though Royal Dutch also exploited its oil monopolistically in the Anglo-Dutch colonies and other areas of the world, neither Holland nor England suffered this exploitation at home.

While Standard was able to monopolize the United States market, Royal Dutch tried to get in wherever possible. Standard exercised its control by manipulating prices at whim, whereas Royal Dutch cornered the greatest number of wells. Deterding's great logistical capacity drove him into the United States market via the State of Washington. He there founded the Gravenhage Association, a corporation in which Royal Dutch, as well as

---

[3]Harvey O'Connor, *World Crisis in Oil,* First Spanish Edition, Ediciones y Distribuciones Aurora, Caracas, 1962.

Rothschild, Lane, and Gulbenkian, participated. They proceeded to finance La Corona in Mexico and Roxana Petroleum in the United States Middle West. In 1913 they took over the companies mentioned as well as Gulf Oil.

It was then that Deterding decided to accept United States participation in these companies, as it "was irritating to see a successful company in any country, without the collaboration of nationals." It is curious, and regrettable, that this fair and wise thought of the oil magnate was not applied in the case of less developed exporting nations.[4]

Early in the First World War, Royal Dutch became powerful in North America, where it owned lands in Long Beach that proved to be the richest oil area in California. It had likewise extended operations to all corners of the British Empire where there was oil. In 1911 it took over Rothschild's shares in Manzout Pnito (Russia). Later Royal Dutch acquired General Asphalt, which—through its affiliate, the Caribbean Petroleum Company—held concessions in one of the most productive oil regions of the world: Venezuela.

In 1913, Royal Dutch obtained new concessions in Venezuela for its affiliates, Venezuelan Oil Concessions, Ltd., and the Colón Development Company, on fabulous terms. Another British company, British Controlled Oilfield, controlled by the British government, obtained the Planas concession in 1917. In the meantime, Standard, which already had a refinery in Cuba, set up another company in Brazil.[5] These big oil companies obtained their greatest profits during the First World War. They thrived on that conflict. Oil became indispensable, as the war was already being won "with oil and blood."

From this brief review of the oil scene at the beginning of the century, it is obvious that two fundamental facts made the birth and development of Venezuela's oil industry propitious at that time.

In the first place, there was an ever-growing market for oil products in the industrial nations of Europe and in the United States. The First World War clearly established the dependency of the war machinery manufactured by developed countries on the oil which now gushed forth from the soil of certain nations in greater quantities than in peacetime.

In the second place, the two powerful companies, Standard and Royal Dutch, were already in a stage of intensive development. Great competition and rivalry erupted between them, driving them to eagerly seek new

---

[4]It is just now, in the 1970s, that this criterion is being widely applied by exporting nations, as a result of pressure exercised by their governments—today members of the powerful Organization of Petroleum Exporting Countries (OPEC), founded in 1960.

[5]Standard of New Jersey established its first affiliate in Venezuela, Standard Oil of Venezuela, in 1921.

sources of oil. By 1913 Royal Dutch had already penetrated the Western Hemisphere, in the United States and Mexico. Its next step would logically be Venezuela.

Although Standard was already operating at full capacity in the United States and Mexico, it was unable to fully cover the demands of its international markets during the war. In 1917, near the end of the conflict (the period in which I have placed the beginning of Venezuela's oil industry for the purposes of this book), Royal Dutch exceeded Standard's production by 57,000 barrels per day. It controlled 96 percent of production in the Dutch Indies and 11 percent in Russia, and even its production in the United States was greater than that of Standard. Furthermore, it monopolized production in British Borneo and in Egypt as well as Venezuela's incipient production.

Thus, the contemporary oil industry, born in the middle of the last century, had spread throughout the world by the beginning of the twentieth century. New techniques were created for surface exploration, production, drilling, refining, marketing, and transportation. Powerful oil companies developed; vast markets opened up, fanning the companies drive for power and control over production centers and markets.

The private petroleum companies played an outstanding role in world economic and technological development, and particularly in that of Venezuela. They placed their capital and technological know-how at the service of the industry. Using their own resources, they financed the initial investments when petroleum exploration involved major risks and when international finance institutions were nonexistent. They created a vast technology concordant with requirements of the petroleum industry proper and with the peculiarities prevailing in each producing country. For example, offshore exploration and exploitation of petroleum was developed for the first time in Venezuela, and this event permitted the advancement of deep-sea petroleum exploitation in other areas, such as the Gulf of Mexico and the North Sea.

At the same time, and uncoordinated with these international developments, many structural changes of a legislative, administrative, political, social, and economic nature were occurring in Venezuela. They were to lay the foundation for developing the immense wealth hidden for centuries in the incredible Maracaibo Basin.

In the oil industry (more than in any other economic activity), if we are able to read between the lines of both local and international economic, social, and political aspects, we can see how all these are closely related, harmonizing in a logical and fascinating pattern. The wheels of history synchronized a series of scattered world events into a complex combination, spurring the development of rich oil areas throughout the world.

This was the process at work in the years preceding the buildup of Venezuela's oil industry in 1917, and this was the fundamental international and local environment which set the stage for Venezuela's oil history.

THE VENEZUELAN PANORAMA

Social and Political Background

Venezuela declared independence from Spain in 1811. However, it was not until 1821 that Simón Bolívar's army finally overthrew the Spanish forces at the battle of Carabobo, thus securing Venezuela's era of freedom.

Between 1819 and 1830, Venezuela was part of the republic of Colombia, historically known as Great Colombia, from which it seceded in 1830. At that time the political organization of the present republic of Venezuela took place, as a state independent both from Spain and from Great Colombia.

Throughout the last seventy years of the nineteenth century, the country was subjected to constant uprisings and civil wars. Military dictatorial governments prevailed, except when civilian dictators were temporarily backed by these governments. Thus political power supported by the Armed Forces delayed the advent of a democratic governmental system by more than one hundred years. However, a social democracy was emerging in Venezuela and is in full force today, based on its inhabitants' predominant spirit of equality. Whites, Negroes, and Indians blended to create its hybrid population, in which the ethnic factors forming it became integrated, eliminating racial discrimination. This social democracy has become strengthened, and the advantages of political, social, and economic improvements are gradually reaching all sectors of the population.

In the first thirty-five years of the twentieth century, Venezuela was submitted to two primitive and backward military dictatorships: that of Gen. Cipriano Castro (1899–1908) and that of Gen. Juan Vicente Gómez (1908–1935). Quite by chance, the promotion of Venezuela's petroleum industry took place during the latter dictatorship.

During the seventy-five years of this century, Venezuela has been subjected to military governments for fifty-five years and to civilian and democratic governments for twenty years. In the years of democracy, the last sixteen are significant in the country's institutional life. From 1935 to 1945, the governments presided by Generals Eleázar López Contreras

(1935–1941) and Isaías Medina Angarita (1941–1945) represent a period of transition from absolutist and militarist dictatorial regimes to a liberal and democratic system.

In the transition to Venezuela's political democracy, many facts and dates are signficant. It has been said that Venezuela's modern state was incubated in 1936, because then the struggle for structuring political parties and labor unions commenced, aiming at public freedom and organization of the state based on the people's wishes expressed through suffrage. However, although in practice these postulations began to become effective in 1945, as a result of the October 18 revolution, it has only been since January 1958, when the Marcos Pérez Jiménez dictatorship of eight years was overthrown, that the country finally became stabilized within a democratic system. Thereafter, and uninterruptedly, four constitutional periods succeeded each other, presided over by civilians chosen through free elections: Rómulo Betancourt (1959–1964); Raúl Leoni (1964–1969); Rafael Caldera (1969–1974); and Carlos Andrés Pérez (1974–1979). The institutionalization and neutralism of the armed forces during these periods also maintained a free play of opinion through the political parties.

These elements, deeply ingrained in Venezuela's present political life, are extremely significant in the light of Venezuela's past political reality, which was characterized by the predominance of force, military dictatorship, and autocracy.

Factors contributing to Venezuela's democratic system are the capability of the nation's political leaders; the strength of the political parties in the face of major changes in national government; and the institutionalization and independence of the armed forces throughout this period. In support of the foregoing is the history of the political parties, which were subjected to extremely destructive persecution throughout the dictatorial periods, from 1948 to 1958. Some were declared illegal and their leaders were exiled or imprisoned; others had their most elemental activities restricted. In spite of this persecution, these political parties had sufficient moral strength and courage to reorganize within a period of merely ten months, after 1958, and to regain their power through popular vote, thereby definitely instituting a democratic system.

On the other hand, governmental, private, and political Venezuelan leaders were sufficiently flexible to propitiate mutual agreements and join forces in achieving their common goals in national affairs, thereby contributing positively to strengthening the democratic system prevailing today. All Venezuelans have clearly demonstrated their democratic vocation, and the essence of the representative system established since 1945 became evident when the majority pronounced their will through the 1959 presidential elections.

Venezuela's political and social democracy today represents the most solid system in Latin America and possibly even throughout the Third World. However, where the system still openly fails is in its administration, which is so alarmingly inefficient notwithstanding the ever-increasing oil bonanza. Thus great social and economic problems still overwhelm a high proportion of the population. Therefore an administrative revolution should take place without further delay.

After this bird's-eye view of Venezuela's political antecedents and trajectory up to the present, I shall now refer to the legal and administrative background of Venezuela's petroleum industry proper.

Legislation

Venezuela's oil legislative history began with the ordinances of the Spanish crown and continued with the mining codes and special hydrocarbon laws decreed by the nation after it obtained independence. The first legal structure enforced in Venezuela referring to oil was the Mining Ordinance for New Spain (Mexico), drawn up by King Charles III of Spain in Aranjuez in 1783. It was applied to Venezuela in the following year by royal order.

Article I, Disposition 22 of this ordinance states: " . . . and I likewise concede that discoveries, applications, registrations and denunciations may be treated in the aforementioned manner not only as regards gold and silver mines, but likewise in the case of deposits of precious stones, copper, lead, tin, quicksilver, antimony, smithsonite, bismuth, rock salt and any other fossil, be they perfect metals or partly mineral, bitumens or juices of the earth. . . ." Granted, this text does not mention "petroleum" or "hydrocarbons," but it does contain the generic expressions by which they have traditionally been known and defined, i.e., "fossils, bitumens or juices of the earth. . . ."

This ordinance, dating from the colonial era, was enforced in the republic of Great Colombia[6] by decree dated October 1829; it was then

---

[6]Great Colombia (Gran Colombia) was the republic founded by Simón Bolívar made up of the territories which today form the present republics of Colombia, Panama, Ecuador, and Venezuela. Its name was Colombia, but I here refer to it as Great Colombia in order to differentiate it from the present republic of Colombia. The republic of Great Colombia was created by the Congress of Angostura (which met in the city of Angostura, today called Ciudad Bolívar, in Venezuela) through the fusion of the Union of Nueva Granada (today Columbia) and Venezuela. Its constitution was ratified by the Congress of Cúcuta in 1821. Ecuador joined these two countries in the new republic in 1822. In 1830 the union was dissolved, its members recovering their status as sovereign states.

The union covered an area of 918,181 square miles and was divided into three departments: Cundinamarca (today Colombia), Quito (today Ecuador), and Venezuela. Bolívar ratified by the Venezuelan Congress in 1832 and was in force until superseded by the 1854 mining code. The basis of the ordinance is the principle whereby ownership of mines and hydrocarbon deposits was attributed first to the Spanish crown and later to the Venezuelan state after the nation declared its independence (1811) and after the separation of Venezuela from Great Colombia (1830). This legal principle is still in force.

During the colonial era property owners in Venezuela were also the owners of hydrocarbons found on their land, including oil. This modality prevailed until the New Spain Mining Ordinance came into being in 1783, as there had been no previous ruling of any kind regarding ownership rights of such minerals for the Spanish crown.

However, through the enforcement of the ordinance, and its sanction by Simón Bolívar's[7] decree of October 1829, hydrocarbon-deposit owner-

---

wished to build Great Colombia into a South American state strong enough to counteract the United States on this continent. The United States was at the time already a political and economic power of international influence, and Bolívar visualized its even greater future.

Unfortunately, this project of farsighted political vision could not be sustained over the years due to conflicting regional and personal interests in the provinces or departments which constituted Great Colombia. Its leaders were unable to comprehend Simón Bolívar's foresight. Should it have survived, Great Colombia would today be a state having a population of over 36 million inhabitants, situated in the northern part of South America, with an extended coastline on both the Caribbean Sea and the Pacific Ocean; it would be a true counterweight in international politics, not only to the United States, but to the South American colossus—Brazil, one of the largest nations of the earth. Great Colombia would eventually have exercised powerful world influence, once its economic potential had been developed.

Looking back on that Great Colombia, destroyed by its own members, certain modern-day developments offer striking paradoxes. The present republic of Colombia, which was so influential in the dismemberment of Great Colombia through pressures exercised against Venezuela, is now eagerly seeking economic integration with Venezuela in the Andean Regional Market, formed by the previous members of Great Colombia, in addition to Chile and Peru. Its former sister nations in the Great Colombia union are now keenly interested in Venezuela's outstanding economic position as an oil producer.

[7]Simón Bolívar, born in Caracas on July 14, 1783, is a universal figure whose life and achievements project beyond the ideas and circumstances of his time, placing him among the great leaders of humanity. It is difficult to convey the image of this man in a brief summary, especially as he was the forger of a whole new world. As a thinker, politician, author, warrior, statesman, and internationalist, he successfully stands up to any critical scrutiny, even considered in the light of twentieth-century political and social doctrines.

The emancipation of South America from Spanish colonialism and its political organization are the fruits of Bolívar's actions as a warrior and statesman. The inter-American system encompassing the three Americas (North, Central, and South) was created by his profound political thought, which led him to outline ideals and specific aims for such a regional system. The Congress of Panama, convened by Bolívar in 1826, and his idea of creating a Society of ship rights passed to the Spanish crown and later to the Venezuelan state, which preserved Spanish mining legislation through 1854, the year in which the first Venezuelan mining code became law. The statutes of the code likewise attribute ownership rights to the state. This principle still prevails in Venezuela, though it has been subject to variations because of the nation's troubled political life.

Two opposing political ideas struggled for primacy in Venezuela during the second half of the nineteenth century: one strove for the establishment of a centralist regime, the other for a federal type of government wherein the states, though ceding some of their powers to the central government, would nevertheless retain autonomy in many areas. This conflict of ideas is reflected in Venezuela's mining laws, especially in concepts of mine and hydrocarbon ownership.

When the Federal War[8] (1858–1863) was won by the advocates of the federal principle, Venezuela's Constitution, dated April 22, 1864, granted the various federal entities ownership of the mines in their respective territories as well as the right to administer them. (Previously ownership and administration had been attributes of the national government.) This principle remained in force until 1881. Authorized by the Constitution of 1864, and subsequently by that of 1874, the states of the union started benefiting from their mining ownership and administration rights. In practice, this led to a confrontation between the federal and state governments, because the latter began making their own mining codes and granting mining concessions.

American Nations, are concrete manifestations of the inter-American system. Besides liberating a continent, Bolívar contributed to establishing the basis of its political organization. The constitutions of Venezuela (1819) and Bolivia (1826) are outstanding examples of Latin American constitutional rights; they are based on Bolívar's thought and ideology.

Over and above Bolívar's liberation of many nations, and regardless of the conflicting realities of their social and political transformation, he bequeathed an extensive and sage body of political thought that might well serve as the basis for a doctrine making feasible the democratic state which is today seeking its form and content in all parts of Latin America.

To absorb and follow many of Bolívar's teachings gives us keen insights; a study of these teachings will help us sow conscientiousness along the route humanity will walk in search of its most treasured principles: the self-determination of nations, individual liberty, and democracy. I believe that in Bolívar we might find many appropriate answers to the anxieties of our time, for the doctrines that are America's very own are rooted in Bolívar's thought.

[8]From 1858 to 1863 a civil conflict known as the Federal War shook Venezuela. It originated in the different outlooks dividing groups and political parties participating in national life. The political outcome of the war was the institution of a federal regime; socially it marked a new start in the struggle to achieve equal rights for all sectors of the population. Multiple gains have been made in this area in the past century since, and social equality is today firmly established in Venezuela.

This anomalous situation led to the reform of the mine-ownership principle in the 1881 Constitution, also federalist-oriented, and to the introduction of a system whereby ownership of mines is still attributed to the states of the union, but their administration is reserved to the federal government. This dual system of regional ownership and centralized administration established by the 1881 Constitution was maintained in subsequent Venezuelan legislation. Variations in mine ownership and administration rights since 1854 may be summarized as follows:

1. The central government entitled to ownership and administrative rights (1854 to 1864).

2. Federal states where the mines were located held ownership and administrative rights (1864 to 1881).

3. Ownership held by federal states where the mines are situated, and administrative rights by central government (1881 to the present).

It should be observed that the ownership granted to the states of the union since 1864 has been merely symbolic, because it has been the central government which has always exercised the use and enjoyed the benefits of the nation's oil deposits.

To grant ownership of mines to federal states and administration rights to the central government was a legislative contradiction of the federal regime. If in 1881 there were political reasons for keeping up the prestige of the states in relation to mines, it is no less than absurd that in subsequent constitutions—including that of 1961, which is today the law of the land—this anachronistic principle continued in force. It is regrettable that ownership rights are still classified as the property of the individual states when these do not, in fact, exercise such rights. This principle should disappear from Venezuela's legislation.

The fact that Venezuela's central government—and not the states of the union where the hydrocarbon deposits are situated—is the recipient of the country's oil income is considered unfair by many. For this reason, and as a compensatory measure, the 1961 Constitution presently in force establishes the possiblity of making "special economic allotments favoring states where hydrocarbon mines are situated." This regulation has not yet been applied, but it will eventually be implemented because of its underlying principle of *equity*. The Bolívar District in the state of Zulia has yielded more of Venezuela's oil wealth than any other area; nevertheless, its community needs have traditionally been neglected.

In the historical evolution since 1783 of mining rights attributed to the state, the 1915 mining code is remarkably significant. Article 85 of that code establishes the inalienability of oil deposits, asphalt, bitumen, and pitch. Since then, presidential powers to grant concessions for the exploitation of these substances have been restricted.

All the foregoing concerning mining ownership has a legal and economic significance of outstanding importance, because in that ownership lies the difference between the oil history of Venezuela and that of other Latin American countries. Unlike Venezuela, Mexico and Colombia eliminated the so-called royalistic system of the New Spain ordinances from their legislation, and instead adopted the accession system (the landowner is the proprietor of the deposits). Because of this legal disposition, Mexico's takeover of its petroleum began with the nationalization of the deposits belonging to the landowners and continued with nationalization of the oil companies. In Venezuela, however, nationalization can be conceived of only in relation to the industry in all its phases of exploitation, simply because the nation's oil deposits have always belonged to the state—with Venezuelan law having always incorporated this basic principle of the New Spain ordinances.

The procedure whereby private individuals could acquire rights for exploration and exploitation of oil deposits has also varied according to the laws which have superseded each other in Venezuela. The state granted the right initially under the system of denouncement (typical of mining law) and later through special contracts and concessions. From the first 1854 mining code to that of 1904, the right to explore and to exploit hydrocarbons could be acquired by means of the denouncement system, as this legal procedure was commonly employed in the case of metal and hydrocarbon mines. Any citizen, be he a Venezuelan or a foreigner, was entitled to make a denouncement, and the state was obliged to grant him exclusive rights to explore and exploit the hydrocarbon deposits specified, with no restrictions except those stipulated in the mining code. These were, then, obligatory grants.

In the laws which superseded the 1904 mining code, the denouncement system was abandoned, and the state was empowered to grant rights to explore and exploit hydrocarbons through special contracts or to deny application for such rights. This state-grant system for hydrocarbons first appeared in the common regime of mining codes and rules long before the common regime was made distinct from general mining law and the first separate hydrocarbon law drawn up. The optional grant system was preserved in subsequent laws, with slight variations, first as special contracts up to the 1921 law, and then as concessions from 1922.

The terms "special contracts" and "concessions" led to discussions within the legal field concerning the legality of existing hydrocarbon concessions granted by the state to private individuals. Discussion revolved around whether these concessions were contracts subject to common law, or whether they had a special status governed by administrative law. I mention this in passing, as the legal aspects of such governmental contractual modalities and their effects are beyond the scope of this book.

Administrative Antecedents

Regarding economic and political events prior to the advent of Venezuela's oil industry, I merely wish to emphasize certain antecedents I feel are of interest, as they will serve as points of reference for analyzing the development of Venezuela's oil industry throughout its several phases.

The first ten-year-period concession for exploiting hydrocarbons was granted to Camilo Ferrand in August 1865 by the state of Zulia and covered the entire state. In February 1866, the New Andalucía state legislature granted a twenty-five-year concession covering all its territory (at present Sucre and Monagas states) to Manuel Olavarría.

Those grants were a result of the restoration in 1864 of the federal political regime, whereunder the provinces, or states, became autonomous entities, and as such, were entitled to legislate and make decisions in mining matters. Exercising this privilege, several states of the union enacted their own mining codes, among them the New Andalucía Sovereign State Code in 1866 and that of the state of Bolívar in 1881. After 1881, when a new Venezuelan Constitution was passed, the administration of mines was delegated to the central government, and henceforth the mines were exploited under a uniform system which has been in force ever since.

In 1878, Manuel Antonio Pulido obtained a concession covering 247 acres close to Rubio, in the state of Táchira, for a fifty-year period. Pulido founded Petrolia del Táchira Company, which has the distinction of being the first oil producer in Venezuela. The company obtained its oil from the upper formation known as Mito Juan situated in the La Alquitrana field, close to San Cristóbal. This company was a strictly Venezuelan entity founded and operated by an illustrious Venezuelan, a pioneer in the field. It obtained one of the first government-granted concessions and operated for more than half a century (up to 1934) in all phases of the industry. It is regrettable that this company was unable to pave the way that would have changed the immediate destiny of Venezuela: systematic, progressive, and regular participation by private Venezuelan entrepreneurs in the oil business.

In the case of Petrolia del Táchira, it is thought that its slight projection in the Venezuelan oil industry was due to the fact that the La Alquitrana zone which it exploited was neither commercially profitable nor potentially rich in oil. Royal Dutch-Shell later also tried to exploit La Alquitrana, but it did not achieve favorable results either. On the other hand, it is true that the establishment in Venezuela years later of the world's major oil-exporting industry would require great amounts of capital and organization and technical skills not available to the Venezuelan private sector, or even to the Venezuelan state.

But it is nonetheless regrettable to see that early participation by Venezuelans in the industry was reduced to the easy, and in general not too honorable, line taken up by some in the first years of the industry, of trafficking in concessions. These concessions, which rendered juicy profits to a handful of government officials and private individuals in the era of Cipriano Castro[9] and more particularly in that of Juan Vicente Gómez,[10] represented the gratuitous handout of national wealth to a very small group of Venezuelans. They also signified enormous profits for foreign companies, which enjoyed the benefits of these vast concessions granted on excessively advantageous terms.

With rare exceptions, such as Pulido's Petrolia del Táchira, participation in the industry by Venezuelans from the private sector has been limited first to dealing in concessions and later to rendering services to the concessionaires. As far as the government is concerned, its only incursion into management of the industry has been, since 1961, the promotion and administration of the Venezuelan Petroleum Corporation (CVP), which in 1973 produced an average of 82,500 barrels per day, representing less than 3 percent of that year's total production. At present the government is scheduling total nationalization of the industry in 1975.

Continuing this brief reference to certain economic and political administrative aspects of Venezuela's oil industry, I might add that in 1883 the government granted a concession for exploiting the asphalt of Lake Guanoco, situated in the state of Sucre. The beneficiaries were Horacio R. Hamilton and Jorge A. Philips, who later transferred the rights to the New York & Bermudez Co., a subsidiary of General Asphalt, a United States firm.

Thus, at the end of the nineteenth century, foreign investors became interested in obtaining concessions in Venezuela and also—as would be the case some years later in our own century—in interfering, in one way or another, in the nation's internal affairs. The case of General Asphalt, accused of having helped finance Gen. Manuel Antonio Matos's unsuccessful "liberation revolution," is an example.

This state of affairs gave rise to two claims by Venezuela against New York & Bermudez Co. The first, dated July 20, 1904, sought to obtain dissolution of the contract drawn up in 1883 between the Ministry of Development and Horacio R. Hamilton, plus payment for damages aris-

---

[9]The government of Gen. Cipriano Castro (1899–1908) emerged from a successful uprising. It subsequently degenerated into a dictatorship.

[10]The government of Gen. Juan Vicente Gómez took over as a result of the coup d'etat that overthrew Gen. Cipriano Castro on Dec. 19, 1908, opening the door to a new dictatorial regime which ruled up to 1935. This regime was the most prolonged and cruel dictatorship ever known by Venezuela. During Gómez's reign, Venezuela's oil industry was coincidentally started and developed.

ing from its nonexecution. The second, dated September 22, 1904, was based on the company's promotion and support of Matos's revolution. Both claims were granted, the first on August 7, 1905, the second on June 27, 1908.[11] Subsequently, Gómez's regime restored the concession to its titleholder, and it was exploited up to 1934. This meddling by a foreign company in Venezuelan political life, originally directed toward overthrowing Castro, contributed to solidifying Gómez's position.

Venezuelan governments started handing out oil contracts during the first decade of this century. During Cipriano Castro's and Juan Vicente Gómez's regimes the government entered into special contracts covering hydrocarbons which are the most highly criticized in Venezuela's oil history. That criticism is based on the territorial extensions involved, the extremely advantageous conditions granted the concessionaires, and the slight or nonexistent benefits reaped by Venezuela. Notable among these contracts are the following.

On January 31, 1907, a concession was granted to Andrés Jorge Vigas for a fifty-year period covering an area of nearly 5 million acres in the state of Zulia. It was transferred by its titleholder to the Colón Development Company, a subsidiary of Royal Dutch-Shell.

Antonio Aranguren was granted a concession (February 28, 1907) for exploiting asphalt in the Bolívar and Maracaibo Districts; it was later transferred (October 1913) to the Venezuelan Oil Concessions Co., a subsidiary of Royal Dutch-Shell.

Francisco Jiménez Arraiz received a concession (May 18 and July 3, 1907) for exploiting asphalt, other substances, and oil, covering 1,235,522 acres in the states of Falcón and Lara. It was then transferred to the North Venezuelan Petroleum Co., a subsidiary of Anglo Persian, controlled by the British government.

General Bernabé Planas was granted a concession (July 22, 1907) for exploiting oil, asphalt, and other substances in the state of Falcón. In July 1909 it was transferred to British Controlled Oilfields.

Rafael Max Valladares, a well-known Venezuelan lawyer, received an oil concession (July 1910) covering an unspecified area in the east of Venezuela, which he transferred four days later to the Bermudez Company. The Bermudez Company, a subsidiary of North American Asphalt Company, was frequently and erroneously referred to as Bermudez Asphalt. In January 1912, this same Valladares—an agent of General Asphalt—received another concession which, incredibly, covered twelve states of the republic (Anzoátegui, Carabobo, Táchira, Monagas, Mérida,

[11]C. E. Thurber, *The Origin of North American Capital in Venezuela,* trans. Angel Raul Villasana, Editorial Nueva Segovia, Barquisimeto, 1955, pp. 66 and 195.

Lara, Trujillo, Yaracuy, Territorio Delta Amacuro, and part of the states of Zulia, Falcón, and Sucre). Years later, in 1920, its area was estimated at 67 million acres.

From these concessions it can be seen that by 1912 a large area of Venezuela's territory already belonged to foreign companies through the transfer of long-term contracts, subject to low taxation and with hardly any participation by Venezuela in profits derived from exploitation. Further, the contracts granted export-tax exemptions. Judging from these facts, we see that Venezuelan beneficiaries of concessions openly aimed at transferring their exploration and exploitation rights to foreign companies, such negotiations being benevolently looked upon by the Venezuelan government in those days, notwithstanding opposition and warnings voiced at the time. J. M. Spindola, Mining Director of the Ministry of Development, opposed R. M. Valladares's contract, and Dr. Leopoldo Baptista, a member of the government, forewarned that it would be transferred. It was nonetheless approved by the government, only to be transferred two days later.

The special contracts or concessions gave rise to controversies between the Venezuelan government and private individuals who had lawsuits brought against them. This led to the adoption of measures limiting presidential power to enter into special hydrocarbon contracts, which now had to be submitted to Congress for approval and were subject to dispositions established in regulations and special laws concerning coal, oil, and hydrocarbons (1915 and 1920). These regulations established the basic terms for all future contracts relative to hydrocarbons granted by the Venezuelan government, as well as the principle of inalienability, whereunder hydrocarbon deposits cannot be sold by the government.

The early contracts or concessions, covering more than 74 million acres, which brought the nation hardly any benefit but which did make millionaires out of their holders, later passed mainly into the hands of the Caribbean Petroleum Company, a subsidiary of Royal Dutch-Shell.

In July 1914, Caribbean Petroleum started regularly producing 250 barrels per day from the Zumaque Well in the Mene Grande field located in the eastern Lake Maracaibo zone, originally part of Valladares's concession; this oil discovery, fundamentally important to the development of Venezuela's petroleum industry, was made in the oil-rich Lake Maracaibo Basin.

Concessions of such great proportions and incredibly advantageous terms, censurable from all moral points of view, did, however, succeed— by sheer chance—in arousing the interest of one of the world's three largest oil companies: Royal Dutch-Shell.

This circumstance, combined with the tremendous wealth of Venezuela's oil deposits (dramatically emphasized in 1922 by the blowout in Shell's No. 2 Los Barrosos Well in La Rosa field, where oil gushed at the rate of 100,000 barrels per day for nine days), prompted development of the industry, with multiple effects. Another colossus—Standard of New Jersey—was attracted, as well as other companies such as Gulf, thus leading to the development within a few years of the most powerful group of international companies ever known in the history of humanity. Some people maintain that Gen. Juan Vicente Gómez deserves credit because, through his proliferous handout of concessions to relatives and friends, he facilitated the first stage of the oil industry's development, i.e., the exploration of Venezuela's territory in search of hydrocarbons. I do not share that opinion. That may have been the end result of Gómez's actions, but it was certainly not his aim. His purpose was entirely the opposite and totally unpatriotic, for he merely wished to illegitimately enrich himself and his friends and to deliver the country into the hands of foreign capital in order to cement his domestic and foreign political power.

This is the same kind of reasoning used by some Venezuelan sociologists in trying to justify the roles of Cipriano Castro and Juan Vicente Gómez in Venezuela's political life. It is said that both dictators contributed to the total integration of Venezuela by incorporating the Andean states in the heretofore isolated western part of the country, and especially Táchira, into national life, thereby safeguarding and consolidating Venezuela's territory.

I reiterate that though this may have been the result of that political phase, it was certainly not a conscious purpose of its protagonists. Even if such outcomes are accepted at face value, they cannot, and must not, be used to attenuate or defend what these individuals actually represented in the history of Venezuela—a long period of reactionary dictatorship. Even if the concessions granted during Gómez's government favored exploration of Venezuela's territory and the establishment of its oil industry, Gómez's procedure in these vital matters cannot be justified, as it was totally devoid of any desire to protect Venezuela's interests. The unlimited handout of Venezuelan territory to foreign companies during the Gómez era may have facilitated exploration, but it also brought about the nation's marked economic dependency.

Another argument used in defending Gómez's government in regard to oil is the supposition that Venezuelans and their government during the initial stages of the industry were unaware of the fact that the oil industry should be operated for the benefit of the country. This assertion is not strictly true. In Juan Vicente Gómez's government there were men who were, in fact, conscious of what oil exploitation really meant to the nation's economic well-being and who fully realized the regime's mistakes; they pointed them out concretely and unequivocally. For example, I quote some paragraphs from a report by the committee in charge of studying the proposed 1922 law of hydrocarbons and other combustible minerals:

> Oil, coal and other substances are the nerve of modern industry and indispensable elements for the life of civilized peoples; without them, not even their independence is safe, because the armies and fleets of our day cannot operate without an abundant supply of these precious substances. Therefore, nations such as ours which possess these substances, must conscientiously regulate this area of public wealth.
>
> Taking into account differences existing between our country, which is merely a producer, and others, such as the United States, which are also consumers, the criteria of the Legislator must also be different. It is conceivable that in the United States an easy yardstick is used in taxing oil companies, for the country urgently requires oil for a variety of industries that later contribute to increasing public revenue. In Venezuela, where oil is mainly extracted for exportation, it is essential that taxes on exploitation be greater in order to ensure benefits to the nation from production. . . .
>
> The Legislator is confronted by two problems relating to this Project: on the one hand he must guarantee those legally engaged in mining investigation that, after complying with the legalities involved, they can expect to obtain the concession sought without fear of malicious obstruction, or of being subjected to the capriciousness of any public official; on the other hand, he must assure the nation that all its mineral wealth will not be monopolized to the detriment of future generations and the future destiny of the Nation.

Finally, the report mentions that "reserved areas may only be placed under contract following prior survey and no party may allege any right whatsoever to securing concessions therein. Further, in all contracts arising from exploitation permits, lots shall be reserved to the Nation, and such areas constitute excellent future reserves." Some of these concepts are still in force even today.

Thus begins—simultaneously with the development of the companies' international power, and helping to nourish it—the incredible, unexpected, unique, and amazing chapter in Venezuela's history involving an industry which transformed the state into a powerful entity and which enriched a few of its citizens, as it scattered part of the country's ever-increasing income throughout the land. Venezuelan governments, both dictatorial and democratic, have lightly invested, and above all thoughtlessly spent, that fabulous income.

Oil revenues fostered advanced infrastructure in many aspects, especially in certain areas; modernization of some urban centers, especially the capital city of Caracas; the accumulation of millions of dollars in international reserves and in the treasury; and for a time, the importation of all types of products from all over the world, whether essential or not, many for consumption by limited sectors of the population. Increased oil revenues made possible the establishment of ultramodern assembly factories, from which a small part of the population makes a living and a minimal percentage has become enriched, thanks to total governmental protection, which many classify as a privilege granted a small group of industrial impresarios.

Unfortunately—and possibly because of that wealth so easily come by— Venezuela has been unable to create the basic and permanent institutions that would allow its future generations to enjoy the benefits of the oil manna, to which they are as much entitled as the Venezuelans of today. The nation has been unable to spark a legitimate revolution aimed at achieving equitable distribution of this wealth. Such a revolution would transform Venezuela into a country with a healthy and educated population, with adequate housing and employment; into an organized nation, nourished by solid agricultural and livestock industries able to compete internationally. It would also generate a technological patrimony, resulting from many years of directed investment and research.

All in all, a revolution of efficiency is called for, especially at governmental levels, where the administration of the nation's plentiful oil resources has always been handled. It should be a peaceful, patriotic, and enterprising revolution, the fruits of which future generations of Venezuelans should inherit as the rightful legacy willed them by the oil generations of the twentieth century. This is the type of revolution that Venezuela needs and has not yet been able to bring about—a true and legitimate revolution with a healthy and genuine nationalistic content. That is why the Venezuelan people await immediate action from the leaders of their country.

TWO

Formation of the Industry (1917–1935)

For several reasons I feel it appropriate to set the starting point of Venezuela's oil industry at 1917.

In the first place, official uninterrupted statistics reflecting the role played by petroleum in Venezuela's economy have been available only since 1917. During 1918 oil exports appeared for the first time in national statistics; on a volume of 121,116 barrels, income was approximately $190,275. These figures are indisputable evidence that Venezuela's oil industry had become a reality.

In the second place, Venezuela's hydrocarbon industry had become established and was active in exports, transportation, refining, and domestic marketing. In 1917 oil exports commenced on a regular basis, transforming Venezuela thereafter into an essentially petroleum-exporting nation.

In 1917 Venezuela's first oil refinery was established (San Lorenzo) and its first pipeline laid between that refinery and the rich Mene Grande fields (Lake Maracaibo). Both installations were owned by the Caribbean Petroleum Co., a subsidiary of the Royal Dutch-Shell group. For the first time in Venezuela's history gasoline, kerosene, crude oil, and products covering needs of the entire market proceeded from a refinery located on Venezuelan soil.

By 1917 Venezuela's oil industry embraced the main activities essential to its formation. Although not on full scale, established companies had already explored, transported, refined, and sold Venezuela's petroleum abroad and locally. It can therefore be said that the conjunction of these events in 1917 marks the starting point of Venezuela's oil industry.

The petroleum era was born. Its effects upon legislation, taxable income, and the Venezuelan economy were to be felt in subsequent years and up to the present time.

I have established 1935 as the end of this first cycle, because it was then that the fundamental exploratory phase of Venezuelan territory had been completed and the main regions of hydrocarbon deposits discovered; because the oil industry was established and Venezuela had become internationally prominent as a producer and exporter; and finally, because substantial changes in Venezuela's political, economic, and social history occurred after 1935.

In subsequent chapters it will be seen why events after 1935 were oriented toward defining and channeling a current of public opinion to pursue these objectives: strengthening the state's position toward the hydrocarbon-exploiting companies; increasing Venezuela's oil revenue; improving labor conditions; and finally, attaining direct state participation in oil-exploitation activities. The industry would change in aspect after 1935. Formed and developed as it was, its future activity would be characterized by expansion and decline, as will be seen.

INTERNATIONAL ASPECTS—1917

The World Situation

Demand for crude had markedly increased as a result of industrial progress in general, particularly in the automobile industry, and also because of war-machinery needs during World War I. Since then, its usefulness and economic value has transformed petroleum into a front-line source of energy in peace and in war.

In 1917 international oil companies were already entrenched with their powerful technical and economic know-how and were avidly searching for new fields throughout the world to obtain as many concessions as possible. Standard of New Jersey, Royal Dutch-Shell, and Anglo-Iranian were in those days the "Big Three" international companies monopolizing the world oil industry. Throughout those years they, especially Standard of New Jersey and Royal Dutch-Shell, competed tenaciously for production sources and markets.

In 1917 world production was 1,377,784 barrels per day. Main producing countries were: United States (918,674 b/d), Russia (180,321 b/d), Mexico (151,488 b/d), Indonesia (36,110 b/d), Pakistan (22,134 b/d), Persia (19,580 b/d), Poland (17,063 b/d), and Romania (10,195 b/d). Other nations producing less than 10,000 b/d were: Japan (7,838 b/d), Peru

(7,060 b/d), Trinidad (4,389 b/d), Argentina (3,300 b/d), Egypt (2,584 b/d), and Germany (1,759 b/d). British Borneo and Canada also produced oil, although in unimportant quantities.

In this group of producing countries Venezuela held seventeenth place, its initial production being 329 barrels per day. Among nations in the Middle East, on the Persian Gulf, only Persia figured in world statistics (sixth place). Iraq and Bahrain were added in 1927 and 1933, respectively. Saudi Arabia, Kuwait, and Qatar did not begin producing until years after this first stage of the industry's formation.

Therefore the United States, Russia, and Mexico were the world's three great producers in 1917. By 1935 Mexico had dropped to seventh place and the first three became the United States, Russia, and Venezuela.

The oil industry's world panorama during this initial period is summarized as follows:

In Russia, the advent of the Communist regime eliminated this important petroliferous area from the influence of the great international companies, especially Royal Dutch-Shell.

In Mexico, where production had begun in 1901, political and legal changes took place causing the oil companies to move their activities and investments to other countries, among them Venezuela.

In the Middle East, as Persia was the only oil-producing country, this rich area was not fully participating in world production.

Increase in world petroleum consumption and intense rivalry among international companies in their search for new sources, plus the situation in Mexico (adverse to oil companies established there, whereas Venezuela's geographic location and political system were considered advantageous), were the main external factors leading to the formation of Venezuela's oil industry.

Together with these factors favoring the formation of Venezuela's oil industry, other events exercised a negative influence and prevented its major development: the installation of refineries in Curaçao and Aruba and the discovery of other oil fields in the state of Texas (U.S.A.).[1]

The Curaçao and Aruba Refineries

Facing Venezuela's coast, at a distance of approximately 35 miles, are the small islands of Curaçao and Aruba,[2] ancient colonial possessions of Holland. They are part of the so-called Dutch West Indies, autonomous

---

[1]Oil exploitation in Texas started in 1901 through the Spindletop Well discovery. The Gulf and Texas Companies emerged and became the most prominent and powerful.

[2]Curaçao: capital, Willemstad; area, 159 square miles: population, 137,000. Aruba: capital, Oranjestad; area, 70 square miles; population, 59,500.

political realms of the Netherlands. Parallel to the birth and development of Venezuela's oil industry (1917–1929), great refineries belonging to Royal Dutch-Shell, Standard of New Jersey, and the Lago Petroleum Company were established on the islands. The establishment of these refineries had a definite bearing on the development of Venezuela's fledgling industry, since a substantial proportion of its oil would thereby be processed outside its territorial borders.

Both the Curaçao and Aruba refineries, which today process close to one-third of Venezuela's crude, were established for the purpose of receiving Venezuelan oil. Regardless of that obvious aim and its foreseeable negative influence on the development of Venezuela's oil industry, Gen. Juan Vicente Gómez's erroneous policy encouraged, or in any case allowed, the establishment of these refineries outside Venezuelan territory. This deprived Venezuela of a series of economic, social, and fiscal advantages through investments, salaries, and taxes which were attained by the Dutch West Indies rather than Venezuela, the source of their business.

Venezuela's development in refining, a most important phase of the oil industry, was set back. Once these great refineries were installed on the islands, companies extracting crude in Venezuela had no peremptory need to refine locally, and what should have been a natural expansion in Venezuela had to be claimed years later through company agreements and legal reforms. As a consequence Venezuela's refining industry did not gain major importance until years after World War II.

Negative factors brought about by the establishment of refineries outside Venezuela include: decrease in flow of investments; assets which will not be subject to nationalization; and fundamentally, the danger of the refineries' ceasing to process Venezuelan crude or of being used for crudes proceeding from other areas or competitive countries.

This latter problem is already being faced by Venezuela's oil industry. Since 1971 the Shell group has brought increasing quantities of crude from the Eastern Hemisphere (Libya and Nigeria) to the Curaçao refineries, which for logical and historical reasons, were supposed to process Venezuelan oil. These refineries now receive close to 20 percent in competitive crudes proceeding from that area. It is absurd to even imagine the displacement tonnage of tankers bringing millions of barrels of crude from such far-out zones as Africa, to be processed at the very doors of Venezuela, merely a few miles from the large production centers in Venezuela's western area.[3]

---

[3]For example, what would producing countries of the Persian Gulf think, and how would they react, if for specific business reasons, however logical they might be, Venezuelan crudes were processed at the Bahrain and Shuaiba refineries?

As well as retarding the development of refining in Venezuela, the installation of refineries in Curaçao and Aruba reduced the possibilities for establishing a refinery of major proportions on Venezuelan soil for quite some years. In fact, after the Second World War great consumer nations—and even poor nonproducing import countries—began installing their own refineries for processing their increased oil imports.

In addition to practical and technical advantages favoring removal of the refining process from production to consumer centers, numerous economic factors—such as the profits and employment derived from this phase of the oil industry—enabled anxious consumer nations to successfully develop this policy.

I feel this clearly illustrates how Venezuela was affected by a deed, or omission, that in a black moment of its history prevented the establishment of refineries in its territory instead of in the Dutch West Indies. It stands to reason that from the very beginning these refineries should have been built in Venezuela, at the source of the raw material, and even more so because Royal Dutch-Shell, Standard of New Jersey, and Lago Petroleum Corporation, owners of the Curaçao and Aruba refineries, monopolized Venezuela's oil production.

Shell had started operating a commercial refinery at San Lorenzo (Zulia state, Venezuela) in keeping with the world policy then prevailing, i.e., refining should be done at producing centers. Nevertheless, in 1916 Shell installed the Curaçao refinery.

I shall now attempt to analyze these complicated and controversial aspects.

Reasons for and against the Curaçao and Aruba Refineries

From Shell's point of view Curaçao made sense. On the one hand it was Dutch territory similar to where the great Royal Dutch Company was born and where an important sector of its shareholders belonged, thereby securing its investment. On the other hand, the Lake Maracaibo shores offered the disadvantage of navigational difficulties through the undredged Maracaibo Bar. This problem could have been technically solved even then, however, as permanent bar dredging and also pipelines were feasible. Furthermore, the Paraguaná Peninsula coast, as proved later, offered no obstacles and could clearly have been selected as a refining center instead of Curaçao and thereafter Aruba.

A clearly conscientious government would have required Shell to locate its great refinery on firm land in the Paraguaná Peninsula, close to the oil wells. This would have represented economic and practical advantages, i.e., pipelines for transporting crude over short mileage from the wells to the peninsula instead of small coastal tankers needed to carry it to the Dutch islands.

The first argument, that Shell was seeking maximum security for its investments through the Dutch colonies compared to Venezuela's traditional coup de'états, still remained. General Gómez had been in power for only nine years and was yet unable to prove that that phase had ended. Gómez had personal political reasons (aimed at securing his power) for not wanting development in distant and somewhat isolated regions of Venezuela such as Maracaibo or Paraguaná. He feared losing his grip by not being close enough to intercept, repress, or neutralize insurrections, labor upheavals, and even separatist movements.

On the other hand, war threats used by industrial nations to intimidate poor and puny Latin American countries and other less developed continents are well known. Fresh in Gómez's mind was the bombing of Puerto Cabello by German and English ships at the beginning of the century because his predecessor, Gen. Cipriano Castro, was reluctant in paying Venezuela's debts. Possibly General Gómez thought Great Britain or Holland would eventually send its ships to Lake Maracaibo under pretext of protecting oil produced by Royal Dutch-Shell. This violation of Venezuela's sovereignty could also come from the United States, calling to mind former President Rómulo Betancourt's citing of Sumner Welles's (U.S. Under-Secretary of State) 1943 declaration on the Monroe Doctrine and its application in the first decade of this century: "Many American republics could not be called sovereign, because their sovereignty was susceptible to being violated by the United States."[4]

So it was that General Gómez's personal ambitions, plus the conservative tendency of the Dutch, both in Europe and in their colonies, were influential factors in the choice of Curaçao for refining.

Toward 1916, when the new refinery was being planned, Curaçao was going through an unemployment crisis. At that time the Phosphate Mining Co. had closed down, increasing unemployment. Flocks of workers were emigrating to Venezuela, Panama, and Santo Domingo, so the Dutch authorities intelligently welcomed the project. In 1914 Curaçao's metropolitan government had modified its legislation concerning exports and internal consumer taxes, for the purpose of favoring its oil industry.

In the following paragraphs I shall refer separately to some aspects of the early history of the establishment and development of oil refineries in the Dutch West Indies.

Curaçao

During 1914–1916, agreements were drawn up between the Dutch government and the Bataafse Petroleum Maatschappij N.V. (BPM), a subsidiary of Royal Dutch, for establishing a refinery in Curaçao, and on

---

[4]*Venezuela: Policy and Petroleum*, Senderos Editorial, Caracas, 1967.

October 20, 1916, the Curaçaosche Petroleum Maatschappij N. V. (CPM) was registered in The Hague. In 1915 the BPM parent company had bought the land for the refinery in Curaçao, and the first two storage tanks were installed in March 1916. Although installations were not completed until some years later (1923), the plant started operating in 1918. First exports of Venezuelan crude to the refinery consisted of 6,300 barrels transported in wooden tankers.

After the First World War (1920), Europe's consumption of gasoline increased. Transportation between Venezuela and Curaçao was more efficient, and two distillery units were installed, plus a conversion unit. During the 1920s workers migrated to Curaçao from neighboring zones. Venezuela provided specialized labor; Holland sent employees of medium and higher categories. Nonspecialized workers emigrated from Bonaire, Aruba, Trinidad, Antigua, other Caribbean British islands, and even Portugal. The ethnic majority predominant in Curaçao today was formed. In smaller numbers, Lebanese, Chinese, and Jews from Europe arrived, definitely contributing—especially the latter—to the island's boom.

Obviously, this important development, starting in the 1920s and, with ups and downs, lasting to the present, is primarily the result of a refinery which should have been established in Venezuela, where in addition to employment benefits and commercial activity, taxable income from refining and imports would have been forthcoming. Venezuela yielded all these advantages to Curaçao by passively allowing the installation on the island of a plant with great refining capacity for Venezuelan crude.

Products refined in Curaçao were logically exported mainly to Europe, as the plant belonged to Europe's biggest company. During many years Europe absorbed the bulk of Curaçao's exports, thus becoming Venezuela's first oil market, until Venezuelan crude and Curaçao's refined products became dependent on the United States market. Due to several circumstances (for example, the competition of Middle East and North African oil), Venezuela's position in the European market has shrunk in recent years, to occupy a secondary place.

Aruba

Aruba's refinery was established in the San Nicolas area (western Aruba) and started production during the second half of 1919. Its history and development is somewhat different from that of Curaçao's refinery due to prevailing international oil circumstances, and especially because Aruba represented interests of United States companies, i.e., Royal Dutch-Shell competitors.

The establishment of Royal Dutch's Curaçao refinery marked a precedent followed by its oil colleagues and competitors already operating in

Maracaibo in the 1920s. These were the Dominion Gulf (today Mene Grande) and Indiana Standard, which later sold its Venezuelan properties to Standard of New Jersey. Both Gulf and Indiana handled Venezuelan crude during the 1920s at their refineries on the Gulf of Mexico; however, they also sent Venezuelan crudes to the Curaçao and Tampico (Mexico) refineries because of marketing difficulties at that time in the United States.

This circumstance influenced United States companies in deciding to establish a refinery in Aruba, another island dependent on Holland, close to Venezuela's production centers.

Even in the face of a second major refining plant on Venezuela's doorstep, dictator Gómez's attitude was the same (as it remained until his death): discouraging refineries in Venezuelan regions far from his political domain, such as the states of Zulia and Falcón, or in any event, allowing their location outside Venezuela.

By the time the Aruba refinery was established, United States oil interests were entrenched in Venezuela, competing with Royal Dutch in sales of refined products to the domestic market, with Standard using a small refinery it had built in La Salina (Zulia state). However, a period of harmony and practical cooperation soon began between the giant United States and European groups.

Royal Dutch had also installed a small refinery in Aruba, close to San Nicolas, operated by a subsidiary, the Eagle Petroleum Corporation. This plant was the first to export refined products from Aruba, but it was not greatly expanded. Since it was not producing aviation gasoline, it closed down during the Second World War, and its personnel was absorbed by the Curaçao Shell refinery, then operating at full capacity. Finally, in 1952, the Eagle refinery in Aruba ceased altogether, and its assets were transferred to the Curaçao Shell refinery and to the Aruba Lago refinery.

In reference to United States penetration in Aruba, I shall briefly mention competition between Royal Dutch and the United States companies, especially Standard.

Relations between the Dutch and United States government were sometimes tense because of this competition, and these incidents are a revealing index of the decided support big United States and European companies had from their respective governments, particularly in the first stages of the industry and in the period when their tremendous power was created. Standard was fully backed by the United States government in its overseas operations, especially after 1911 when Standard of Ohio was legally forced to dissolve and Standard Oil of New Jersey took over management of the Rockefeller group of companies.

Around 1920 to 1921 the United States and Holland were involved in diplomatic bickerings, because the United States was pressing to enter exploitation of the Dutch East Indies oil fields. While United States companies had to overcome a series of obstacles and restrictions by the Dutch government to be able to get in, Royal Dutch received vast concessions in the Djambi area in 1921. Because of this, the United States government claimed the Dutch government had violated the Open Door policy, which according to the United States government, originated from the 1920 Dutch mineral concessions law. In retaliation the United States refused authorization to Royal Dutch for operating in Utah.

Pressure exercised by the United States throughout the decade in protecting United States companies' oil interests—especially Standard's—prompted the Dutch government, fearing new United States retaliatory measures, to sign contracts with United States companies in 1928, enabling them to engage in oil exploitation activities in Indonesia. Possibly this situation influenced Royal Dutch and Deterding—interested in entering the United States oil industry—to persuade the Dutch government to allow the establishment of a United States company-controlled refinery on colonial Dutch territory.

In this shrewd game of world oil chess, lasting throughout the history of the industry, it is enlightening to see how the opening up of Indonesia's petroleum area to United States companies contributed to the establishment of Aruba's vast refinery.

Finally, let us now see what this refinery represented, and still represents, to the economy of Aruba.

Just like Curaçao, Aruba was in a precarious economic situation prior to the establishment of its refinery. Its small population depended on agriculture and livestock, primitively and rudimentarily developed. Employment in Aruba, as in Curaçao, was extremely limited because of its barren land, lack of water, and scarcity of markets. In the face of such poverty, the establishment of oil refineries in both Curaçao and Aruba was more than a blessing, but the benefactor had a name and surname: "Venezuelan Oil Wealth," unprotected and unsafeguarded, to be sure, by Venezuela's leaders in those days.

REFINING IN VENEZUELA

At this point it will be well to briefly describe refining activity within Venezuela during these years.

As crude oil has no direct marketing use, it inevitably has to be submitted to physical and chemical refining processes in order to obtain final consumer products. This refining process is what gives oil its extraordinary commercial value.

Venezuelan legislation regulated refining (taking it as just another phase of the oil industry, such as transportation) beginning with the 1922 law of hydrocarbons, which established types of concessions for each activity integrating the oil industry: exploration, exploitation, manufacture or refining, and transportation.

Where refining was concerned, it was merely a concession complementary to exploitation of autonomous concessions. In the first case it consisted of a right inherent to a concessionaire for exploitation, which he was entitled to use after complying with the requirements of the law of hydrocarbons and during the validity of his original concession, i.e., forty years. Autonomous refining concessions had a fifty-year validity.

During 1917 to 1935 oil companies were not legally or contractually obliged to refine in Venezuela. Venezuela's government lacked a refining policy, and if oil companies established plants for covering domestic requirements, this was due to their own policies aimed at monopolizing the local market and obtaining tax exemptions.

Refining in Venezuela started with the type of activity developed by Petrolia del Táchira Company, owned by Manuel Antonio Pulido, which in 1880 installed the first refinery with a capacity of 20 barrels per day of kerosene in Rubio, Táchira state. In 1900 the Val de Travers Company installed a small refinery in Delta Amacuro, and in 1910, in the Guanoco pitch lake (state of Sucre). However, both were for refining asphalt.

Due to its greater capacity, the first commercial refinery established in Venezuela was that in San Lorenzo, installed on the eastern shore of Lake Maracaibo (Baralt District in the state of Zulia). This refinery, owned by the Caribbean Petroleum Company, started operating in August 1917, processing crudes coming from the Mene Grande field, to which it was joined by a 10-mile pipeline. Its initial capacity was 2,000 barrels per day, and by 1926 it was 10,000 barrels per day.

At the end of this phase there were eight refineries in Venezuela producing 16,800 barrels per day, representing only 6.1 percent of its crude production. Through legal modifications, new concessions after 1935 included—as a special advantage to the nation—the obligation by concessionaires to refine a higher percentage of crude within Venezuela.

THE ACHNACARRY AGREEMENT

In the 1920s, particularly in 1928 and coinciding with the Aruba refinery installation, important events happened in the world petroleum industry. In 1928 the famous Achnacarry Agreement was drawn up between the big oil companies (Royal Dutch, Standard, and Anglo-Iranian), ending their tough competition and opening a period of basic agreement and cooperation framed in moderate world competition. From that moment, relations among the big companies improved considerably, and it is worth giving certain details of this historic agreement.

On September 17, 1928, a meeting was held in Achnacarry Castle, in Scotland, by Sir Henri Deterding, Walter C. Teagle, and Johnson Dadman, representatives, respectively, of Royal Dutch-Shell, Standard Oil Company of New Jersey, and Anglo-Iranian—the Big Three—to discuss problems affecting the oil industry, or better said, affecting the companies they represented. What has since become known as the Achnacarry Agreement[5] emerged from this meeting.

The agreement is one of the most remarkable events in oil history. The situation prevailing in the world petroleum industry at that time, and the future objectives of the companies, conscious of their economic power, are clearly reflected in its principles and provisions. Excess oil production since 1925 had caused an outbreak of a price war between Royal Dutch-Shell and the Standard Oil Company of New Jersey, especially in India and Great Britain, affecting the interests of both companies. The purpose of the Achnacarry Agreement was to put an end to that situation and stabilize markets. It recommended that (1) consumption requirements of a specific geographical area should be covered by its own production; (2) production should be curbed in regions where surplus occurs; and (3) future production should be controlled and oil prices established, based on Gulf of Mexico reference prices, regardless of the product's point of origin.

The agreement originated joint action by the companies in confronting problems arising from their own competition, and it secured their firm control over the industry, which they were to hold another thirty-two years, up to the 1960 constitution of the Organization of Petroleum Exporting Countries (OPEC).

The Achnacarry Agreement is divided into three parts: a preliminary justification of measures adopted, an announcement of seven principles, and fifteen provisions of practical application. Because of their importance to world oil history, I transcribe hereunder, almost literally, the seven principles and the most significant of the fifteen practical provisions set down in 1928:

Principles

1. To accept each group's present business volume, which will be a reference basis for calculating future increases.

2. To make existing oil refining and treatment installations available to producers to avoid useless and costly duplication of refineries, the

---

[5]The official name of the Achnacarry Agreement: "Pool Association of 17th September 1928." See Christopher Tugendhat, *Oil: The Biggest Business,* Editorial Alianza, Madrid, 1968, p. 125.

producer to pay a price at least equal to the cost paid by the owner of the refinery, and maximum equal to the producer's cost had he built his own refinery.

3. To desist from any type of complementary facility unessential to meeting demand.

4. To acknowledge the advantage in covering consumption needs of a geographical area by supplies produced therein.

5. To secure maximum transport savings.

6. To reduce production in excess areas, or to offer such surplus to other markets at competitive prices.

7. To discourage, in the public and private interest, any measure increasing costs and consequently reducing consumption.

Practical Provisions

1. The agreement does not apply to United States imports or exports. These will be covered by a special agreement.

2. Quotas will be calculated on a six-month basis, for establishing each group's rights in each country and in the world market group.

3. Oil prices will be fixed at each world point in relation to Gulf of Mexico reference prices, regardless of the oil's origin.

4. Private agreements will establish quality standards so as to standardize products.

5. Oil fleet surplus, not used by owner groups, will be pool-auctioned.

6. Products will be swapped to economize transportation costs.

7. Surplus production will be offered to members of the agreement at a lower price than to outsiders.

Point 1 shows that the United States market was excluded from the provisions relating to imports because of the United States antitrust law; however, its exports—then the greatest in the world—would be especially controlled by administrative organizations created for executing the agreement.

In practice the Achnacarry Agreement was executed by two committees formed by representatives of each member: one committee in New York, for production control, especially United States production, and another in London, for coordinating measures applicable to consumer markets.

In the United States the antitrust export rules were superseded by the Webb Act (1928), whereunder exports would be channeled through a special office. On the other hand, Standard Oil of New Jersey and its group formed the Export Association (1929). This association established sales prices and quotas for each member. However, since it could not control total exports and in 1930 it was denounced and declared illegal by companies not affiliated with Standard of New Jersey, it was unable to meet the Achnacarry Agreement principles.

Adapting the Achnacarry Agreement to their own purposes, the big companies later drew up other mutual agreements covering market quotas and price regulations (European Market Memorandum, Basis of Distribution Agreement, and Preliminary Memorandum of Principles).[6] These agreements (1930, 1932, and 1934, respectively) not only extended the companies' field of action but expanded their oil cartel, because in addition to the initial parties (Standard Oil of New Jersey, Royal Dutch-Shell, and Anglo-Iranian), they were signed by Gulf, Texas, and Socony Vacuum, with the possibility of independent producers joining in.

I shall not elaborate on these agreements. I merely wish to underline the Achnacarry Agreement's fundamental significance in relation to world petroleum history: it was the first joint action, the first clear-cut agreement between great international companies, for the exclusive purpose of establishing a worldwide oil equilibrium to their own advantage.[7]

To complete the picture of the status of Venezuela's oil industry from 1917 to 1935, I shall conclude this chapter with brief reference to companies then established and functioning and to production figures.

OIL COMPANIES IN VENEZUELA

Venezuela's oil industry has always been, in its totality and with overpowering preponderance, in the hands of foreign companies, especially Anglo-Dutch and North American. Other foreign companies, of Belgian, French, and Spanish origin, as well as Venezuelan, have operated too, but have never become important. In this phase, oil companies in Venezuela were noted for their sustained international and domestic competition through the successive predominance of Anglo-Dutch and North American groups.

Anglo-Dutch companies were the first to develop Venezuela's oil industry, because they were concessionaires of the first hydrocarbon special contracts granted by the government (1907) to the Venezuelan citizens mentioned in Chapter 1. Through these concessions, nearly all areas, covering more than 92 million acres, granted for exploration and exploitation up to 1918 were held by Anglo-Dutch companies: Colón Development Co., Venezuelan Falcón Oil Syndicate Ltd., Minerales Petrolíferos Río Paují, Bermudez Company, and Caribbean Petroleum Company, all belonging to the Royal Dutch-Shell group.

Royal Dutch was a Dutch company founded in 1890 by August Kessler and headed since 1900 by Henri Deterding; it exploited oil in the Dutch East Indies. In 1907 it joined the Shell Transport and Trading Company,

---

[6]Tugendhat, loc. cit.

[7]In Chap. 6 I shall refer to the impossibility in those days for the formation of an association such OPEC among exporting countries.

a British company founded in 1897 by Marcus Samuel, engaged in transporting oil products in the Far East. It thus became Royal Dutch-Shell, its policy being to secure sources of supply throughout the world. Its rival, Standard Oil Company of New Jersey, was oriented to oil exploitation within the United States market.

This contrasting policy of the powerful oil industry's founders, although exhibiting a greater or lesser degree of intelligence in evaluating world petroleum projections, was mainly due to the special geographic conditions prevailing in their backgrounds. Standard, based in the United States, at that time had sufficient sources of production and markets, whereas Royal Dutch and Shell lacked oil in their home territories (Holland and England), obliging them to seek it elsewhere, either in their own colonies or in other independent countries. The industry's development in general, and the First World War's practical lessons, are the factors which finally and conclusively gave the petroleum industry its international character.

The only company fully operating in Venezuela by 1918 was the Caribbean Petroleum Company, of the Royal Dutch-Shell group. According to its 1918 annual report submitted to the Venezuelan government, Caribbean Petroleum produced 922 barrels per day, from 274 declared wells, of which 396 barrels per day were exported and 526 processed at its San Lorenzo refinery. It paid $172,562 to the Venezuelan government in oil taxes in 1918 in addition to general taxes, and on a monthly average it employed 574 persons.

United States companies became interested in Venezuela's oil business in 1919, i.e., after the First World War ended. In 1921 Standard Oil Company of Venezuela, a subsidiary of Standard Oil Company of New Jersey, was established. It first started operating in Lake Maracaibo and later in eastern Venezuela, this area being developed after 1926.

As mentioned, Standard Oil Company was founded by John D. Rockefeller in 1870, in Ohio, with a capital of $1 million. Its assets in 1904 amounted to $110 million. In 1911, a legal decision through the antitrust law ordered its liquidation. It was then split into thirty-five companies controlled by Standard Oil Company of New Jersey, which had been founded in 1882 with a capital of $3 million. Standard was, and still is, a large holding company, owning or controlling many subsidiary companies. It does not directly explore, exploit, refine, transport, or sell, but rather does so through its numerous affiliates operating integratedly throughout the world. After the First World War,[8] Standard started its

---

[8]After the First World War (1914–1918) Standard's expansion outside of the United States was a known fact, acknowledged by its own representatives. The following is attributed to an assistant of Walter Teagle, president of the company: "I believe the future of Standard,

ceaseless drive for power over the world's oil deposits, channeling its investments toward Venezuela, where it became established in 1921.

The Orinoco Oil Company, of Delaware, U.S.A., subsidiary of the Pure Oil Company, was registered in Venezuela in February 1923. By 1926 its concessions covered 294,000 acres.

The Lago Petroleum Corporation was established in Venezuela in June 1923. This company obtained a concession of 116 contracts involving 2,864,000 acres in the prolific Maracaibo Basin, covered by the lake. Lago was the first United States company to export oil from Venezuela. In 1926 it acquired the British Equatorial Oil Company rights. During its first nine years Lago passed through many hands, until May 1932, when it was acquired for Standard of Indiana by Standard Oil of New Jersey, a deal that also included the Aruba refinery. In 1943, most of Standard of New Jersey's interests in Venezuela were merged with Lago Petroleum Corporation under its present name, Creole Petroleum Corporation.

The Richmond Petroleum Company, a subsidiary of Standard Oil Company of California, was established in Venezuela in 1925.

The Venezuelan Pantepec Company was established in the same year, and by 1927 it had selected about 988,500 acres to start an extensive drilling campaign.

The Venezuelan Atlantic Refining Company, of Delaware, U.S.A., was registered in Venezuela in January 1926.

The Texas Petroleum Corporation started operating directly, and through the California Petroleum Corporation, in 1927; in 1931 it renounced its Zulia concessions, and it restarted operations in Monagas in 1939.

More than one hundred companies were established in Venezuela by 1931. This was mainly because most of them—for company policy—were affiliates or subsidiaries of other companies. Later on, and as exploratory and exploitation activities gradually merged, the number of oil companies diminished. By 1932 only eleven companies had reached exploitation stage. They belonged to three powerful international groups: the Royal Dutch-Shell group (Caribbean Petroleum Company, Venezuelan Oil Concessions, Colón Development Company Limited, British Controlled Oilfields, North Venezuelan Petroleum Company, Tocuyo Oilfields of Venezuela Limited, and Central Area Exploitation Company Limited); the Standard group, subsequently Creole Petroleum Corporation (Lago Petroleum Company and Standard Oil Company of Venezuela); and the Gulf group (Venezuelan Gulf Oil Company).

especially Standard of New Jersey, lies beyond the United States." Quoted by Harvey O'Connor, *World Crisis in Oil,* First Spanish Edition, Ediciones y Distribuciones Aurora, Caracas, 1962, p. 66.

The Venezuelan Petroleum Company

The Venezuelan Petroleum Company (Compañía Venezolana de Petróleo, or CVP) was constituted in 1923 by Gen. Juan Vicente Gómez, President of Venezuela (directly or indirectly) from 1908 to 1935. I feel it important to mention this company because it clearly illustrates the political situation prevailing in Venezuela when its oil industry was established: it was the main instrument for oil concession grants made by the government from 1923, and it clearly demonstrates the absence of any petroleum policy by Gen. Juan Vicente Gómez's dictatorial government.

Venezuela has had two companies with the initials CVP—but with opposing origins and purposes—namely, the Venezuelan Petroleum Company, a private company fouunded in 1923 and President Gómez's instrument, to which I am now referring; and the Venezuelan Petroleum Corporation, the present CVP, a state company founded in 1960 and owned by the nation. The first CVP was established for manipulating the nation's oil interests—for granting hydrocarbon concessions and handing out Venezuela's patrimony, regardless of collective interests.

According to Venezuelan general legislation and the hydrocarbon law of 1922, Venezuela's President was not personally entitled to hydrocarbon concessions. To evade this prohibition and participate directly in concession grants, General Gómez created the Venezuelan Petroleum Company. In that way the President received state-granted concessions through an intermediate legal company, which turned out to be his own and which later negotiated with parent oil companies.

To enumerate the hydrocarbon concessions granted to CVP by the government and to describe their area is a task by far too lengthy and intricate for inclusion in this book. However, according to the records of the Ministry of Development (then in charge of the petroleum policy), CVP signed from 1923 to 1931 more than three hundred concession contracts for a total area on the order of 7 million acres. Those concessions were promptly transferred to foreign oil companies, the following among them: American Venezuelan Oilfields, Venalka Oil Company, Venezuelan American Corporation, Misoa Petroleum Company, Sobrantes Pantepec Company, and Venezuelan Seaboard Company.

The second CVP was founded by Presidential Decree No. 260 dated April 19, 1960, clearly for enabling the state to directly exploit its own oil for the nation and not for illegitimate private interests.

Production Figures

Venezuela's peak oil production figure was reached in 1970, showing an average of 3,708,000 barrels per day. Other recent significant figures are:

1 million barrels per day in 1946, 2 million in 1955, and 3 million in 1962. Venezuela's oil production started to be signficant in 1920; year by year, for ten years, it duplicated its respective figures. From 1922 to 1929 the tendency was upward; from 1930 to 1933, the most acute phase of the United States Depression, which spread to the rest of the world, Venezuela's oil production dropped, although keeping an even level of 11.5 million barrels per year; in 1934 and 1935 it again recuperated its rate of increase.

According to official Venezuelan sources, production stopped duplicating from 1930 because of the enormous world production prevailing in those days, especially in the United States and Russia, where it considerably exceeded consumption. This superproduction led to a price drop, and the big companies had to slow down their activities.

By 1935 Venezuela was producing 14,831,631 barrels annually, thereby demonstrating that its oil business was a formed and developed industry in this first stage (1917–1935). Venezuela's accumulated oil production in this period was 1,148,245,699 barrels—in other words, 168,445,605 metric tons. Venezuela's position as an oil producer at that time was significant: from seventeenth place in world statistics in 1917, it rose to second place in 1928, where it remained up to 1931. It was then displaced by the Soviet Union, passing to third place until 1935, the last year of this period, where it remained until 1944. In 1945 it recuperated its second place, until 1960, when it again was displaced by Russia. Thereafter, up to quite recently (1969), Venezuela held third place as a world producer.

With the growth of the hydrocarbon industry during these years, Venezuela's dependency on one single activity controlled by foreign capital, represented by oil companies, was incubated. The part oil revenue played in the formation of its national budget is a clear index of that dependency: Venezuela's public finances gradually and uninterruptedly became more and more subjected to its oil industry. In fact, by 1920 oil revenue was 1.4 percent of national income; by 1932 over 25 percent; and in 1935 over 29 percent. By 1935 Venezuela's oil dependency had reached such a point that it became difficult for the government—under such circumstances—to freely make decisions regarding the petroleum industry without affecting the national economy.

At the end of this period Venezuela had become clearly defined as a developed oil-producing nation, economically dependent upon oil and foreign capital. During the period 1917–1935 Venezuela's fiscal income derived from oil was $92 million on an accumulated production of 1,148 million barrels, excluding an estimated $69 million from import exonerations. To have an idea of Venezuela's very low income level from oil companies throughout this nineteen-year period, we can compare this figure with Venezuela's oil revenue of $8.6 billion received in 1974 alone!

Institutionalization
(1936–1945)

Many changes took place during this period in Venezuela's political, economic, and social life that were greatly influential in the institutionalization of its oil industry. I shall now attempt to pinpoint the most significant.

Through labor and income tax laws Venezuela took a step forward, regulating its social and fiscal aspects in relation to its oil industry.

In general terms, benefits obtained from a specific industry are reflected in the companies it involves, in its personnel, in the community where it is situated, and in the revenue it represents to the National Tax Bureau. This was not the case in Venezuela's oil industry. From its birth to Gen. Juan Vicente Gómez's death in 1935, the companies involved and a few Venezuelans profited enormously, but hardly any benefits were gained by the workers, the communities, or the state.

On the other hand, the industry did represent important intangible contributions, such as the incorporation of managerial and administrative techniques, which contributed to Venezuela's progress. The petroleum companies were greatly influential in this respect, as they trained Venezuelan professionals who, upon being transferred to other economic activities, contributed that acquired know-how to the benefit of private and public sectors.

According to government estimates during the period July 1919 to June 1936, oil companies in Venezuela grossed $1,666 million, on a production of 1,262 million barrels. Oil revenue received by Venezuela's

National Tax Bureau was $118 million for that period. In other words Venezuela's oil revenue from concessionaires was 7 percent.

At the time of dictator Gómez's death, living conditions of the Venezuelan populace were in general deplorable. Unemployment was continuously increasing, reaching high levels difficult to estimate because of lack of statistical data corresponding to that period. In the midst of such a situation, development of the petroleum industry in a certain manner softened the prevailing economic crisis, because its exploration and exploitation activities led to new sources of employment. The oil industry's development created the largest concentration of workers ever known. Armies of men from all over the country, attracted by improved salaries, arrived at oil fields. This, of course, brought along serious social problems.

The workers had reached a stage of such drastic poverty that even the higher oil salaries were insufficient to cover their most basic needs. The situation became so critical that, in spite of the prevailing adverse political climate, a strike broke out in 1925, eleven years before the 1936 strike which most Venezuelans erroneously regard as the first of this type in their oil industry. The workers' predicament persisted relatively unchanged until 1936. Official reports of the times summarize the situation and show that a high number of foreigners were occupying key jobs, enjoying high living standards. Living conditions of native workers were particularly pathetic at Lagunillas and Mene Grande (state of Zulia).

Notwithstanding these negative facts, it is only fair to admit that, since 1936, Venezuelan workers in the petroleum industry have received higher salaries than in other fields of activity. Also, because of the more favorable working conditions in the petroleum industry since 1936, the special labor legislation that was introduced, and the contributions of labor movements, other sectors of national economy became influenced to increase salaries and to improve the economic and social conditions of their respective workers.

When General Gómez's dictatorship ended, one of his principal lieutenants, Gen. Eleázar López Contreras, the Minister of War and Navy, took over. In spite of the troubled atmosphere in which he had to govern, a modern and institutional period began in Venezuela. During his government, serious concern arose over the oil industry. Labor legislation was promoted and a labor movement formed; years later it was to prove its maturity. The legal order was still confusing, but oil companies managed to carry out their activities within an atmosphere of confidence which contributed to speeding up the industry's development.

At the same time an attempt was made to improve the lot of the working class and the Tax Bureau's participation in relation to oil-company profits. This transformation opened the door to legislative reforms and innovations in seeking more equitable solutions; the labor law was passed in 1936; the income tax law in 1942; the hydrocarbons law in 1943, and the Presidential decree in 1945.[1]

The workers, the community, and the Tax Bureau, having their own interests in the oil industry, were the objects of this legislation. I have called this second phase "institutionalization" precisely because of that multiple and intense legislative activity.

The three pillars for future development of Venezuela's petroleum affairs were the labor, hydrocarbons, and income tax laws. These special laws have constituted since then the backbone of Venezuelan oil rights.

1936 LABOR LAW

The 1936 labor law was of great legal and social significance, because it treated labor as a special and autonomous subject within Venezuelan legislation and because it attempted to protect the workers and improve their precarious standard of living.

At the beginning of Gen. López Contreras's[2] term there was a high percentage of unemployment. The need for sources of work originated a series of improvised public works and emergency measures. The government presented to Congress, for urgent approval, a proposed labor law. Its purpose was to regulate worker-employer relations, not merely as a contractual element dismembered from the general body of common rights, but as a social and economic structure entitling workers to certain social benefits implied in work contracts, providing them (and employers) with organizations and instruments for self-defense, and subjecting them to a special labor jurisdiction.

The 1936 labor law accords with universally accepted principles. It establishes an eight-hour working day and a forty-eight-hour working week; compensations for layoff and seniority; profit participation; yearly vacations or rest; freedom to organize labor unions, compulsory social security; and special labor jurisdiction. Although a prior labor law had been in existence since 1928, it did not include any of these special and autonomous dispositions. For this reason the 1936 text should be considered as Venezuela's first labor law.

[1]The purpose of the Presidential decree was to increase state participation in oil companies' excess profits.

[2]Provisional President from Dec. 17, 1935, and then constitutional President to May 5, 1941.

Foreign sources used for the wording of the law include Chile's labor code, Mexico's federal labor law, and labor rulings of Colombia and Peru. Also taken into consideration were international labor agreements signed by Venezuela up to 1933 and recommendations by the International Labor Bureau. Dr. David Blelloch, Labor Inspection Service Chief of the International Labor Bureau created in 1919 by the Versailles Peace Treaty, collaborated in the wording of the law. Venezuela is a founding member of the Bureau.

Although the 1936 law contained many generalities, it seems that its fundamental beneficiaries were oil workers, consisting of an army of more than 12,000 men receiving very low salaries, deriving no social benefits, and having no labor union or legal protection whatsoever. Advantages gained by oil workers under the labor law include an eight-hour workday, prior-notice benefits or compensations, seniority, vacations, and profit participation. But the basic advantage gained was the right to form labor unions and collective contracts, fundamental to the development of all labor legislation. Without including such rights, the 1936 law would have contributed little to the social development that began in Venezuela in those historic years.

Labor unions took their first struggling steps in spite of systematic resistance from the oil companies and an adverse political climate. Years later, through collective contracts developed by the unions, Venezuela's working class started receiving a more reasonable profit participation in the oil industry. I shall refer to this again later.

Oil Workers' Strike

In 1936 Venezuela's oil industry was fully developed. From second place as a producer in world statistics (1928, 1929, and 1930), Venezuela had dropped to third place; however, production had risen from 135,246,418 barrels in 1930 to 154,639,494 barrels in 1936. In other words, while slipping percentagewise in world statistics, its production had increased by close to 7 million barrels per year.

Even more important than production figures is the fact that the industry, which started in Venezuela's western oil basin, was being consolidated in the east. The eastern basin started developing in 1926, and was based on Standard Oil Company of Venezuela's advanced activities in Quiriquire. By 1936, out of 154.6 million barrels produced, 23.3 million were produced in the eastern zone (Monagas, Sucre, and Delta Amacuro) and 131.3 million in the western area (states of Zulia and Falcón). Although production in the western area had always been notably higher than in the east, the east was already an important source of exploitation.

Ship terminal facilities were available in Caripito and a refinery would be established. In other words, the eastern zone already had its own physiognomy in Venezuela's oil industry.

Parallel to this increased development, the number of persons employed by oil companies had also increased, forming the industry's working class. Oil companies employed approximately 12,335 workers by 1935. The political transformation which developed after Juan Vicente Gómez's death was reflected in the working sector; unions were organized, the oil industry's first official one having been founded in February 1936 and called Oil Workers' and Employees' Syndicate of Cabimas (state of Zulia).

In 1936 two labor conflicts broke out in the oil industry: the so-called June march and the December strike. The latter, of major proportions, has become generally known as the 1936 oil strike. On December 11, 1936, oil workers and employees in Cabimas and Lagunillas went on strike in the western area of Venezuela, suspending activities until January 22, 1937, i.e., for forty-three days. This strike was restricted to the western area; eastern-area activities were not affected. The economic effect of this strike was a 39 percent decrease in production, representing $1.3 million which the workers and, of course, the National Treasury failed to receive.

The Strike's Profile

The 1936 oil strike was fair and legal, supported by public opinion, and in principle, by labor administrative authorities. Even though this was the oil industry's first organized strike, the union maintained a strong position not only by calling the strike through legal channels but also by remaining calm and nonaggressive while seeking a solution.

The Caracas *La Esfera* newspaper, in its edition dated December 14, 1936, stated: "We have studied the Lagunillas petitions and must say that our workers' demands are fair and their claims have been submitted in a appropriately legal manner. It is to be hoped that oil companies will meet these requests with normal modifications, so that all points that might become a subject of controversy or of conflict are satisfactorily resolved in keeping with our Labor Law."

On the other hand, the oil companies' attitude in confronting the strike was negative toward the unions' legal, economic, and social petitions. Company representatives rejected all petitions, ignoring the unions and refusing to enter into collective contracts, stating that they had received instructions from their principals to make no concessions whatsoever. Furthermore, they gave the government a list of workers' names to encourage police intervention.

The government's reaction was ambiguous and contrary to the workers' interests. On the one hand public administration officers recognized the workers' rights, but on the other Gen. Eleázar López Contreras's government disregarded them, and through Presidential decree dated January 22, 1937, ordered all strikers back to work, basing its decision on the prolonged length of the strike, on the failure in reaching an agreement, and on the alarming social-economic damages it was causing the nation.

A salary increase less than that requested by the workers was granted, and the strike ended. Not only had the government deserted the workers in their fair demands, but it even forgot its own claims which it had made on several occasions to the oil companies due to various infringements upon Venezuelan laws. Finally, it stunted union development by allowing oil companies to ignore unions as workers' representatives and not draw up collective contracts. It was because of this that it took another ten years (1946) before the first labor contract between workers and oil companies operating in Venezuela was drawn up.

HYDROCARBON CONCESSIONS

Special hydrocarbon contracts and concessions were granted in Venezuela under the military governments presided over by Generals Cipriano Castro, Juan Vicente Gómez, Eleázar López Contreras, Isaías Medina Angarita, and Marcos Pérez Jiménez. By contrast, and even symptomatically, none were granted under the constitutional governments of Rómulo Betancourt (1945–1948 and 1959–1964), Rómulo Gallegos (1948), Raúl Leoni (1964–1969), and Rafael Caldera (1969–1974). In Venezuela's current constitutional government, President Carlos Andrés Pérez, besides not granting any further concessions, has clearly announced that the industry is to be nationalized.

General Eleázar López Contreras (1936–1941) adopted two policies in the matter of concessions: initially his government granted concessions covering approximately 2.47 million acres, but later on it ceased completely, right up to when he was succeeded by Gen. Isaías Medina Angarita. Concessions granted by López Contreras's government to Standard Oil Company of Venezuela, Socony Vacuum Oil Company, Ultramar Exploration Company Limited, Compañía Consolidada de Petróleo, and Texas Petroleum Company stipulated—as a special benefit to the nation—that concessionaires would be obliged to refine part of Venezuela's exploited crude in Venezuela, through the expansion of existing refineries and installation of new plants. In compliance with this condition Standard Oil Company of Venezuela inaugurated the Caripito refinery on October 22, 1939, in the eastern area of the country, with a 30,000-barrels-per-day capacity.

Public opinion was critical of these concessions. The Venezuelan private sector thought them to be so vast that the oil companies, even jointly, could not possibly drill the whole area in the following fifty years and that therefore the comapnies' purpose to monopolize Venezuela's oil deposits was quite apparent. Some officials openly opposed the traditional policy of granting concessions, and the Chamber of Deputies' Hydrocarbons Committee as well as the First Inspectors Conference (held in Caracas in 1938) asked the government to not grant any more concessions.

INCOME TAX AND OIL REVENUE

On several occasions unsuccessful attempts were made in Venezuela to establish income taxation. The National Congress received several projects between 1937 and 1941, but the first income tax law was not passed until 1942.

The 1942 income tax law radically changed Venezuela's tax system, because direct taxation was thereby included in national revenue, which until then had depended mainly on indirect taxes. The context of this law is based on recommendations by the North American Economic and Financial Mission, presided over by A. Manuel Fox, and by the Fiscal Committee of the Society of Nations at its meetings in The Hague (April 1940) and Mexico City (June 1940).

Venezuela's oil revenue has always played a major role in its ordinary fiscal income. In 1974 it amounted to $8.6 billion (85 percent of ordinary fiscal income). The 1974 oil income tax represented more than three-fourths of the total oil revenue. Its increase since 1943, when it was applied for the first time as a general taxation, was from $1.3 million in the first year to $588 million in 1960 and then to $8.6 billion in 1974.

State participation in the oil industry's income logically increased significantly when the 1942 income tax law was introduced, because the law created a new source of taxable income by enforcing a general taxation on all income. Modifications in later years increased taxable income due the state from oil companies even further.

Prior to 1943, oil revenue was fundamentally derived from special taxations contained in the existing hydrocarbons law, the most significant being that applicable to exploitation (royalty), representing approximately 60 percent of this income. In 1942 oil production represented approximately $28 million to national fiscal revenue, equivalent to 30.18 percent of the national budget. After the income tax law was established, oil revenue increased for the first year (1943) to $87 million, of which $6 million corresponded to income tax. By 1945 this income tax paid by oil companies had increased to $14 million.

To illustrate the favorable effect participation by the state had in profits proceeding from the oil industry, I might add that in 1961, oil revenue exceeded $393 million. Between 1944 and 1961 the accumulated figure was $3.6 billion. Income tax received in fiscal 1973 was $1.8 billion. In view of these facts, the income tax law's significance to the National Treasury in relation to the oil industry is undeniable, demonstrating that oil reforms after 1943 were accomplished fundamentally through tax laws and not through hydrocarbon laws.

At the beginning of this chapter I mentioned that fiscal elements would be regulated in this second stage of Venezuela's oil history, a fact which has certainly been proved by the country's present income tax and its positive results. The income tax law was created prior to President Medina's so-called oil reform, the 1943 hydrocarbons law, but increase in fiscal income derived from oil started when both laws were applied in 1943.

Oil Legislation Reform—1943

Venezuela's important 1943 hydrocarbons law, approved by Congress, represents the best-conceived and best-structured legal document concerning hydrocarbons that the country has ever had. It responded to a national clamor for a revision of Venezuela's oil policy, and from a political viewpoint, to the government's promise to carry out an oil reform.

During Juan Vicente Gómez's dictatorship, his opponents denounced the government's petroleum policy as negative and detrimental to national interests, pointing out the goals to be achieved in a matter of such utmost importance to the country's economic, social, and fiscal life. In 1931 a group of exiled Venezuelans, in their so-called Barranquilla Plan, censored the Gómez regime for—among other criticisms—its responsibility in the nation's unfavorable oil position. After Gómez's death in 1935, discontent was high and wide over unfair participation in oil profits and the deplorable living conditions of the industry's working class. In spite of the 1936 and 1938 hydrocarbons laws, the need for oil reform persisted.

Although the succession of laws passed from 1920 to 1938 might well have represented progress within the legal structure, they meant very little in relation to national economic interests, because concessionaires disregarded them and defiantly continued operating under the laws prevailing when they obtained their concessions, which were less demanding. Consequently, these new laws merely represented a theoretical, formal expression of Venezuelan rights with no practical meaning, incapable of imposing a general ruling applicable to all concessionaires.

In fact, up to 1942 Venezuela's participation in oil-industry profits was not uniform, because multiple concessions with diverse royalty percentages existed. Royalties realized on general hydrocarbon production in 1942 were split as follows:

32.5% (the older concessions) paid only $0.51 per metric ton
  36% paid a 7½% royalty
 2.7% paid a 8½% royalty
23.8% paid a 10% royalty
   5% paid a 15% royalty
100%

Under these circumstances, Venezuela's participation was extremely unfair, because any tax increase in new laws would mean little if not applicable to all concessionaires. Naturally such a frustrating situation of national interest was surely to breed resentment, and it became a doctrinary banner for the political parties that were being organized at that time and that started participating in national development, especially Acción Democrática.

In the international sphere, and especially in Latin America, powerful public opinion arose, causing governments to adopt legal measures for protecting petroleum resources. Nationalization of Mexico's oil industry—expropriated in 1938 and to which I shall refer later—was then a recent event.

Also influential was the Second World War declared in 1939, with its inevitable economic effects on markets and transportation. The idea and purpose of a revision of Venezuela's oil policy, born as a nationalistic expression before the war as a result of Venezuela's political and economic realities, arose anew with the added circumstances imposed by that international conflict.

General Isaías Medina Angarita's government, whose Minister of Development was Eugenio Mendoza, introduced the oil reform that its opponents—and the country—had been insistently seeking. This governmental initiative, apart from its specific finality, was politically motivated to combat the main objections of the opposition. As is generally the case with laws, regardless of the positive aspects quite rightly attributed to the 1943 oil reform by its partisans, this law did not fully satisfy national interests, and several sensible criticisms were made when it was debated in Congress. Nevertheless, the law was passed on its original terms.

The 1945 Revolution

Many important changes occurred between 1935 and 1945 in Venezuela's political history, representing improvements to the foundations of its institutions. Juan Vicente Gómez's death in 1935 put an end to a regime in which force was the mainspring of political power. Later, under Eleázar López Contreras (1935–1941) and Isaías Medina Angarita (1941–1945), a period of transition began, gradually replacing force by a system of public liberties, although preserving the indirect presidential electoral system.

On October 18, 1945, President Medina's government was overthrown by a joint military-civilian coup. Congress had named Gen. Medina Angarita as President on April 28, 1941, for a five-year period ending in 1946. In the preceding government he had been the Minister of War and Navy (equivalent to today's Minister of Defense). Under General Medina's Presidency a system of full public liberties prevailed and the country was on the way to major economic development. However, political circumstances then existing fomented the October movement which overthrew his government.

The National Constitutional Reform, passed in 1945, did not set forth, as demanded by the people, direct first-degree presidential elections through popular, universal, and secret voting; rather it preserved the second-degree electoral system, i.e., through Congress. The practice of governmental continuance had prevailed in Venezuela, whereby the President of the Republic—as number one elector—was entitled to choose his own presidential successor. General Eleázar López Contreras chose Gen. Isaías Medina Angarita in 1941. In 1945 General Medina had chosen as the official candidate Dr. Angel Biaggini (Minister of Agriculture) for the 1946 elections.

As things stood this practice would persist, because it was obvious that the government would win the 1946 elections; out of 143 voting members in Congress, only 13 belonged to the opposition block. This opposition could do little to stand up against the anachronous electoral mechanism which, preserving the 1945 Constitutional Reform, practically placed election results in the hands of the President through his parliamentary majority.

Venezuela had matured and Gen. Eleázar López Contreras's regime had taken a big step toward constitutionalism, but political institutions had remained stagnant. Possibly a provisional ruling should have been passed under which the 1946 and 1951 presidential elections would go through Congress but succeeding elections would go through popular, direct, universal, and secret voting. This might have prevented General Medina's overthrow as well as proved that this was the true purpose of those who failed to include direct presidential elections in the 1945 Constitution. At least it would have been one argument less for the forgers of the October 18 uprising, who were the Acción Democrática[3] party associated with a group of military officers.

---

[3]Acción Democrática was founded in 1941 and was the first Venezuelan political party in the entire history of Venezuela that managed to become organized as such. It has, ever since, been the nation's majority party, evident by the results it has obtained in all elections in which

The movement that overthrew President Medina became known as the October revolution. It is an episode in Venezuela's political history on which opinions differ. Some defend it as a true revolution; others repudiate it as merely another coup in Venezuela's troubled political life that just happened to break the thread of constitutionality.

Presidential Decree—December 31, 1945

The Government Revolutionary Council, which assumed power on October 18, 1945, enacted Presidential Decree No. 112 on December 31, 1945, introducing an extraordinary and additional taxation on incomes exceeding $259,000 (Bs 800,000). One of the intentions of this decree was to fill a void in Venezuela's fiscal laws, which did not then tax excess profits. Although it is true that Venezuela's tax system was short in such dispositions, the decree's fundamental objective was to tax excess profits of the oil companies so the nation would get a fifty-fifty profit slice from its oil.

The decree in effect fixed a 20 percent tax rate on income exceeding $647,000 (Bs 2 million). Even though this extra tax included mining activities in general, its direct application to oil companies was obvious because they were the only ones earning such excess profits in Venezuela.

Out of the 20,000 taxpayers then existing in Venezuela, only 75 were affected by the 1945 Decree No. 112, and of these 98 percent were oil companies. The decree generated an extra income of more than $30 million (Bs 93 million) for the National Treasury. Although the decree

---

it has participated since 1946, although excluding the 1968 elections when its presidential candidate, Dr. Gonzalo Barrios, was defeated by the Social Christian party's candidate, Dr. Rafael Caldera. With this exception, Acción democrática has won all elections in Venezuela.

It is a popular and multiclass party, having a dense following proceeding from the labor, student, professional, and business sectors. It was initially oriented to the left, but at present it might be classified as center-left. It conceives the revolutionary process as an evolutionary advancement, its doctrine being essentially Venezuelan, nourished on the realities of the country and devoid of any ties to foreign ideologies or international disciplines. In its foremost manifestations, it is classified as nationalistic and anti-imperialistic. In practice, it has played a predominant role in anti-communist and anti-Castroist conflicts. Its history, as that of its country, has been very agitated; it has acted in opposition, in government, clandestinely, and in exile.

A great number of its leaders—at all levels—have been persecuted, imprisoned, or assassinated, or have lived long lengths of time in exile.

When Pérez Jiménez's eight-year dictatorship was overthrown in 1958, Acción Democrática was legally restored; it restructured its political base, including its national and regional leadership teams, over a short span of ten months (February to November 1958) and obtained a major triumph in the 1958 December elections. In the last elections held in December 1973, its presidential candidate, Carlos Andrés Pérez, obtained a majority of more than 500,000 votes over Social Christian party's candidate, his main opponent.

was provisional, it was greatly important, as it immediately became the basis of subsequent income tax reforms, especially in 1946 and 1948. In 1948 tax on excess profits derived from mining activities in general and from oil activities in particular became permanent in Venezuela's tax system.

Repercussion of this decree abroad was influential in orienting Middle East petroleum countries, which used it as a guideline in establishing their own respective oil-profit participations.

OIL IN MEXICO: EXPROPRIATION AND NATIONALIZATION

Reference to Mexico's oil history is essential, because not only was Mexico one of the first countries to reach prominence in the world petroleum industry, but also it was the first to expropriate oil companies. This historic event happened in 1938.

Mexico's oil position in relation to Venezuela's was, up to a certain point, quite the opposite where hydrocarbon mining ownership and economic and fiscal characteristics were concerned. Oil doctrine and legislation in all countries are structured around the fundamental principle of mining ownership. Ownership of petroleum deposits was attributed to the state in countries such as Venezuela, where legislation includes the doctrine of dominion; in countries such as Mexico, where legislation includes the system of ownership by accession, it was attributed to private landowners. Nevertheless, according to Mexican legislation, oil deposits have alternately belonged to the state: 1783–1884 (the New Spain Mining Ordinances were passed in 1783); to private landowners: 1884–1917 (Mexico's first mining code was drawn up in 1884); and then again to the state: 1917 hereto.

Mexico's 1884 mining code, inspired by prevailing principles of economic liberalism, granted to landowners ownership rights to the land proper, including everything above and below it. This change of concept in mining property rights, which passed from the Mexican state to the landowner, was profoundly significant in the development of Mexico's hydrocarbon industry, not only because of the legal conflicts that it engendered and the irreversible principles of laws and rights acquired thereby, but because the oil industry's development remained subjected to multiple contractual ties.

Mexico's 1917 Constitution abandoned the 1884 mining code and reattributed petroleum-deposit ownership to the nation, as a result of the oil industry's importance and the decisive role it played in the First World

War (1914–1918). For security and public utility reasons, most countries throughout the world applied the system attributing ownership of petroleum and other minerals to the state, and Mexico adopted this decision in 1917.

While these legislative changes were taking place, Mexico's oil industry was established and developing under the system of accession. Exploration and exploitation were channeled through contractual agreements with landowners, sometimes through private third parties, and, when involving nationally owned lands, through the government.

This division of mining property obviously diversified the systems applicable to oil companies. In the fiscal area, exploitation on private ground did not generate any sharing by the nation in oil profits. Furthermore, although the state received no profits, oil companies benefited in turn from incentives and exonerations granted to them by petroleum laws simply because those franchises covered hydrocarbon exploitations. Under such circumstances it was by far more advantageous for oil companies to exploit private ground.

Theoretically, oil nationalization encompasses these goals: attributing mining ownership to the state; reserving exploitation of deposits for nationals, excluding foreigners (as in the United States); total or partial state monopoly of the oil industry (as in the Soviet Union); or resorting to expropriation of the industry for the public and social good.

In Venezuela, the nationalistic tendency in petroleum matters has been, up to the present time, based on the following principles: (1) to preserve state ownership over hydrocarbons and other combustible minerals; (2) to achieve higher participation in the oil industry's profits; (3) to secure direct state participation in the oil industry; (4) to reserve for the state a sector of the industry (as was the case with natural gas); and finally, as is happening today, (5) to completely nationalize the industry.

Distinctions by nationality as a means for nationalizing the oil industry have not been made in Venezuela for legal and economic reasons. Venezuela's Constitution and its hydrocarbons laws treat Venezuelans and foreigners equally. From the economic viewpoint, industrial requirements of capital and technology could not be satisfactorily covered by Venezuelan companies.

Venezuelan nationalism has had to be oriented until quite recently toward the economic and practical measures of obtaining a higher profit participation and of taking over ownership of certain oil-related activities such as petrochemicals, the domestic market, and the natural gas industry. However, present conditions are propitious for the total nationalization of the oil industry, a natural course since hydrocarbons have always belonged to the state. I shall come back to this subject later.

Mexico's oil nationalization was the culmination of a process that started with legal reform and ended with expropriation. In Mexico's 1917 Constitution, hydrocarbon deposits became nationalized, and in the 1938 expropriation decree by its President Gen. Lázaro Cárdenas, the industry proper became nationalized. This expropriation of oil companies involved a combination of several factors. It can be interpreted either as the expression of an anti-imperialist feeling or as a purely nationalistic attitude fanned and sponsored by the oil companies' exaggerated speculative behavior.

I can clearly understand these nationalistic sentiments of the Mexican people and can also see the negative political, social, economic, and fiscal factors generated by the oil companies, but I would also say that Gen. Lázaro Cárdenas's government was not an anti-imperialist regime, at least as far as Mexico's petroleum was concerned. I am not passing judgment on Gen. Lázaro Cárdenas's nationalism, but would say that Mexico's oil expropriation was not the doing of a government that arrived in power with the deliberate purpose of nationalizing foreign investments.

When a government of any country introduces measures to defend its national interests vis-à-vis foreign interests, it is generally classified as anti-imperialistic and as being oriented toward the extreme left. Political life in Latin America has been characterized by the development of governments addicted or opposed to foreign interests—especially North American—in their respective territories.

In the group of Latin American nations, Mexico's 1938 regime did not represent an extreme leftist government, nor did it have in mind expropriation of the oil companies, if we take its own declarations as a basis. General Cárdenas himself, in his official speech on March 18, 1938, concerning the expropriation decree, said that the measure was a decision neither wished nor sought for by the government. Therefore, had the oil companies not been so adamant in resisting the dictates of Mexican justice, their expropriation would probably not have taken place, at least not in 1938.

By expropriating, Gen. Lázaro Cárdenas's government was responding to a situation that arose; it was the result of a historic moment but not of a political program. I am discussing these aspects in order to point out that the immediate cause of Mexico's expropriation of oil companies was fundamentally the attitude of the oil companies themselves.

Nationalistic feelings surge not only because a country's leaders wish to control its economic resources, but also as a reaction to certain activities of foreign investors. Nationalism, in confronting foreign capital, is nourished on the following sources: a specifically subjective and patriotic sentiment, an emotional attitude; and a reaction in the face of negative results arising from activities which, with or without reason, are considered unpatriotic.

Essentially, the reason for the nationalization of foreign interests in most countries is the same, namely: the psychological reaction of resentment when the nation confronts the meager social and economic conditions of the majority of its citizens, who have no effective control over their natural resources and no effective participation in the profits derived by foreign interests from the exploitation of those resources.

Any country that has lived through such circumstances has wished, to a higher or lesser degree, to nationalize foreign companies, usually known for not being overly concerned about the people's interests. However, most countries have not been able to avail themselves of the economic, technical, and even political means necessary to expropriate and nationalize oil companies as Mexico did.

In 1938 Mexico's position as an oil-producing country was the following: Oil production had started in 1901; it developed and increased until 1921, when it reached 193 million barrels (530,000 barrels per day). Thereafter it decreased, reaching its lowest level in 1932 (33 million barrels, or 93,000 barrels per day). By 1937, the year preceding expropriation, production was around 47 million barrels (128,000 barrels per day), of which 25 million (a little over 50 percent) consisted of crude oil and derivatives for exportation. Mexico's domestic oil consumption by 1938 amounted to 46 percent of total production.

In other words, Mexico's oil industry could count on foreign and internal markets and also on the certainty of its daily domestic growth, because the country had become industrialized, and, to a certain point, diversified. Toward 1924 local demand for petroleum products represented 12 percent of production; constantly rising, it had reached 46 percent by 1937, the year preceding expropriation. Undoubtedly, expropriation caused the loss of foreign markets and the oil industry's abrupt decline. However, it did not cause its total disappearance, nor did it drastically upset national economy.

Mexico did not fundamentally depend on oil; income proceeding therefrom was only 10 percent of national revenue. Hence, it was not a monoproducing nation like Venezuela, and expropriation did not bring bankruptcy to the National Treasury or disrupt the economy. For that reason in Mexico, more than in any other country, there existed economic circumstances, in addition to those of a political nature, which propitiated expropriation, regardless of whatever adversities might have been forecast.

The Venezuelan situation was, and always has been, different from that of Mexico. Venezuela's fiscal income from oil was 35 percent of the 1938 total and 65 percent by 1946, thus accentuating its status as a monoproducing nation. For this reason, any upset to that income would cause a serious imbalance to Venezuela's economic life, since Venezuela has no alternative or compensatory income for covering any deficit. Even yet Venezuela cannot substitute its oil income with any other.

Venezuela's oil industry depends nearly entirely on exports; its domestic market is a mere 4 percent. Venezuela has no local industry sufficiently capable of absorbing its oil production. Mexico did have a domestic market that contributed in counteracting the closure of international markets, which resulted as an immediate and direct consequence of expropriation.

The foregoing shows Mexico's favorable possibilities for expropriating. It also shows that Venezuela delayed considering total nationalization until quite recently only because of circumstances prevailing in its economy—such as the vast resources required for keeping up its present rate of production so that simultaneously with nationalization it is able to create the necessary instrumentation for revitalizing the industry. However, as we shall see later, Venezuela now has the economic, political, and administrative possibilities to do so.

Referring once more to Mexico, the immediate and related causes of expropriation in the social, economic, fiscal, legal, and political orders are fully contained in the Mexican government's expropriation decree and in President Gen. Lázaro Cárdenas's speech of March 18, 1938. I shall now describe the events leading to Mexico's expropriation.

In 1936 the Oil Workers' Syndicate of the Mexican Republic submitted a Collective Labor Project to the oil companies, applicable to all activities within the industry and initially contemplating a salary increase amounting to 60 million Mexican pesos. Discussions reached a dead end as the companies alleged that the workers' demands exceeded their economic capacity. Consequently an oil strike was called in May 1937.

Because the strike was affecting national economy, the workers called it off but resorted to presenting the labor conflict as a conflict of an economic nature. According to the then-existing labor law, when companies declared economic incapacity to meet workers' demands, either of the parties involved was entitled to submit an economic petition to the Federal Conciliation Board so that it might assign a committee of experts to analyze the companies' financial status and to judge the applicability of the workers' demands. This is what the workers did, and the Federal Conciliation and Arbitration Board designated a committee of experts.

The committee of experts concluded that the oil companies' financial status "should be classified as extraordinarily prosperous and therefore it is possible to assure that, without any detriment whatsoever to their present and future situation, at least in subsequent years, the companies are perfectly able to meet the demands of the Oil Trade Unions of the Mexican Republic, as a minimum, by a yearly amount of approximately 26 million Mexican pesos."[4]

According to prevailing legal rulings, the Federal Conciliation and Arbitration Board rendered a decision based on the opinions of the committee of experts (December 18, 1937). The companies then appealed to the Supreme Court of Justice demanding protection. The Court confirmed the decision of the Federal Conciliation and Arbitration Board (March 1, 1938), sustaining an increase of social benefits amounting to approximately $7.3 million. The companies stated that they were unable to comply with the decision and that their economic condition only allowed them to reiterate their offer to increase 1936 benefits by approximately $6.2 million. The difference was therefore brought down to close to $1 million.

The Oil Workers' Syndicate, in view of the companies' unwillingness to comply with the decision, went on strike (March 18, 1938). That same evening the government officially announced the expropriation of oil companies. The oil companies' intractability toward the verdict pronounced by Mexican justice led the government to apply the expropriation law (November 23, 1936), safeguarding the principle of authority invested in the Mexican government.

Because of the eminently enlightening aspects of these facts, three conclusions may be reached:

1. Expropriation in Mexico was basically the oil companies' own fault. They propitiated their own expropriation. The government, although not essentially thinking of expropriation, assumed the responsibility demanded of it at that time.

2. Within the international sphere, Mexico's expropriation of oil companies brought about its disappearance as an oil exporter, and consequently, Venezuela had one competitor less in the United States oil market.

3. Finally, it is interesting to see that Mexico's expropriation, besides favoring Venezuela by eliminating such an important competitor in the international market, served as a forewarning to those same international companies operating in Venezuela, which later translated into more liberal and comprehensive attitudes by their subsidiaries.

In 1938 more than twenty foreign oil companies were operating in Mexico, mostly affiliates of big international companies. The groups primarily affected were: the Mexican oil company El Aguila, a subsidiary of Royal Dutch-Shell; the Huasteca Petroleum Company, a subsidiary of Standard Oil Co. of New Jersey; the Sinclair group of companies (Consoli-

---

[4]Jesus Silva Herzog, *History of the Expropriation of Oil Companies*, Editorial Libros de Mexico, S.A., Mexico, 1964, p. 86.

dated Oil Corporation, Mexican Sinclair Petroleum Corp., Pierce Oil Company, S.A., Compañía Terminal Lobos, S.A., and Stanford and Co.); and the California Standard Oil Co. of Mexico.

Expropriation Difficulties

The most critical difficulties confronted by Mexico's government as a consequence of the expropriation of international oil companies (Standard Oil Co. of New Jersey, Sinclair, Standard Oil Co. of California, Royal Dutch-Shell) were the following: lack of technical personnel, because the companies left Mexico; transportation crises due to lack of tankers and carrier ships; the boycott against Mexico by international companies, which threatened anyone willing to buy Mexican oil, even the companies producing machinery for the oil industry; negative propaganda campaigns in the international press (*The Lamp, The Atlantic Monthly,* etc.) and in the national press (*El Economista* magazine); and interference through diplomatic channels by the United States and British governments.

Mexican author Jesus Silva Herzog describes the difficulties of those days: "They would not sell us materials; in those first months it was difficult for us to sell; lack of materials and replacements was overcome by Mexican patching skills—so typical of poor people—improvising—with only a few primitive tools—any complicated part produced in a modern plant."[5]

Added to these initial difficulties were those that inevitably arose when the Second World War (1939–1944) broke out. Its logical and immediate consequence for Mexico was the closure of the foreign oil markets it had managed to obtain; German submarines, in their attempt to cut supplies to the allies, sunk five tankers of Mexico's small oil fleet in 1942 and one more in 1944.

Expropriation Costs

According to all settlements, expropriated assets amounted to $165 million. Twenty-five years after the expropriation proper, i.e., March 18, 1963, not one single cent was owed to the oil companies. The oil companies refused to recognize expropriation or even to deal with the government over the amount and liquidation of indemnities. However, this resistence broke down after 1940, and the companies, one by one, signed the respective agreements covering their indemnities.

In October 1940 Sinclair agreed to accept a total of $8,500,000. In 1942 the remaining United States group of companies accepted an indemnity

[5] Ibid.

payable in five yearly quotas at a 3 percent annual interest rate, totaling $20,137,700.84. In August 1947 companies of the British group also signed their respective agreements for a total of $81,250,000 payable in fifteen yearly quotas, the first (September 18, 1948) amounting to $8,689,-257.85 and the remainder in subsequent years. Mexican expropriation had begun in March 1938 and ended satisfactorily with the last payment in August 1962.

United States companies had received a total indemnity amounting to $32,495,991 by 1947, and the Anglo-Dutch companies had received $81,250,000 by 1962, which, plus interest, totaled the aforementioned $165 million. In that fifteen-year period, and using resources generated by the industry proper, Mexico had canceled the debt originated through the 1938 expropriation.

Evaluation of the Mexican Expropriation

Mexico's experience in directly exploiting its own oil industry is certainly fascinating and illustrative; however, my main purpose is to describe the historic deed of expropriation as one of the most far-reaching events in the annals of world oil industry.

Jesús Reyes Heroles, General Director of Petróleos Mexicanos (Pemex), the state company that took over oil exploitation in Mexico after expropriation, on celebrating the thirtieth anniversary of that event in 1968, reported to the President of Mexico: "Petróleos Mexicanos is at present fourteen times larger than in 1938. It has expanded over the years at a real and accumulative annual rate of 9.5 percent."[6]

In 1938, when expropriation took place, Mexico was producing 105,-500 barrels per day. This level was maintained in subsequent years, but decreased in 1942, 1943, and 1944. More than to expropriation, this decline is attributed to the Second World War, which seriously affected Mexico's growing markets: Italy and Japan. In the group of producing nations Mexico's importance declined, but its individual production remained constant throughout the first three years following expropriation, decreasing, as mentioned, only in 1942, 1943, and 1944. The industry, after expropriation and later exploitation by the state, although not increasing significantly in world statistics, at least revealed its capability to continue operating directly under the state.

However, should expropriation not have occurred, it is quite possible that Mexico's oil industry would have reached a higher level of increase, even delaying development of other sources throughout the world. But the Mexican lesson, although considered by many as contrary to its own

[6]*Petroleum Policy,* Imprenta Nuevo Mundo S.A., Mexico, 1968, p. 4.

development, has been valuable in demonstrating that the oil industry can in effect survive big international companies' management and control, even though Mexico's experience cannot be brandished as a model of independent development.

The world petroleum industry is now being debated between those who at present exercise its control and those who aspire to do so in the future. Since its birth more than one hundred years ago, it has been in the hands of big international companies, and producers have been unable to escape from that grasp.

As a road toward transition, the Organization of Petroleum Exporting Countries (OPEC) was developed in 1960. Today the flow of nationalistic ideology, increasing producer countries' participation in their oil profits, is powerfully surging toward a clearly defined goal: direct control of their own oil exploration and exploitation industries, displacing the big companies' dominance. This goal is, of course, feasible, but it must be well thought out, because petroleum countries must first acquire a sense of security in their own possibilities and an unfailing self-confidence.

The creation and operation of state oil companies is oriented toward overcoming all obstacles. In oil-producing countries it is now the general opinion that exploitation is a basic industry for the nation, and the old thesis of acquired rights is ceding to the aforementioned concepts.

Along this line of thought, control over the world petroleum industry could pass into the hands of producing countries, and an important revolutionary transformation would take place in the history of mankind. Takeover of the economic monopoly exercised by international oil companies on the part of producing countries—the petroleum owners— would be a new phase in the economic revolution of the present and future world.

The world's poorest countries are the richest in oil. If control of that wealth should pass from international companies to producer states, a world situation with a higher equilibrium of forces would arise, due to the increased dependency that developed nations have on oil. Such a movement could even influence the control over other resources and raw materials by Third World countries, and this in turn would contribute to a further equilibrium and to a higher degree of international parity.

Furthermore, in this new frame of affairs big international companies could play another important role, that of contractors (service companies, for example), being, as they are, so totally integrated in the greatest industry ever known by mankind, which has given meaning and content to the economic and historic phase that we are living in: the petroleum era.

FOUR _____

Expansion (1945–1958)

INTERNATIONAL ASPECTS

In the preceding chapters I have discussed two stages of Venezuela's oil industry: its formation and its institutionalization. I have also appraised the most outstanding events occurring in the oil world between 1917 and 1945, which greatly influenced the development of Venezuela's hydrocarbon industry. Intense industrial expansion began around June 1945 in Venezuelan and world oil history, and this marks the beginning of a third stage in Venezuela's oil industry.

In the period following the Second World War, international markets expanded because of the increased demand for crude and derivatives. The reconstruction of countries devastated by the war and their return to peaceful industrial activities demanded a higher consumption of oil. Expansion of the United States market—already a fact by the time the United States joined the war—became intensified in the postwar ear.

Years earlier petroleum had begun replacing coal as the world's main source of energy. In 1913 coal represented 90 percent and oil 6 percent; in subsequent years, oil spiraled, reaching 32 percent in 1957 (the last year under this appraisal). See Table 4-1.

Because of its industrial growth, the United States, although the world's major producer, required more and more oil and thus became also the world's major importer. In 1945 its domestic needs were 5,358,000 barrels per day, and by 1957 they had increased to 9,386,000 barrels per day, resulting in an average annual increase of 4.8 percent.

This increased demand exceeded the domestic supply in the United States, and therefore the deficit (equivalent to 12 percent) had to be counterbalanced by imports proceeding from other areas, especially Venezuela and Canada. During this period other world areas emerged as producers: Saudi Arabia figures in world statistics since 1946, Kuwait also, and Qatar since 1950. When these new countries entered the oil world, production increased tremendously: from 7,109,000 barrels per day in 1945 to 17,639,000 barrels per day in 1957.

After the Second World War, Russia returned to the scene in many facets: exploration, discovery of new processes, and exports. Furthermore, new companies entered the international panorama, such as Arabian Oil Company (Japanese), Ente Nazionale Idrocarburi, ENI (Italian), and the Bureau de Recherches de Pétrole (French).

During this stage of incipient nationalism, international oil companies entered producing countries because of greater benefits granted and because of their contracting procedures, presented in the form of an association with the producers. Refining—which until then was located close to producing centers—started shifting to consuming countries. Formerly, two-thirds of refining had been developed within producing countries or in neighboring areas. However, the situation has reversed and more than two-thirds of crude is now refined in consumer countries.

The reasons for this shift in refining centers are obvious. Refining of crude in consumer countries provides several economic, technical, and even political advantages: for example, the creation of new sources of employment; savings in freights and currency; control over domestic supplies; and product distribution simplification. In Chapter 6 this subject is covered more fully.

In the twelve years after the Second World War, other international events took place, inevitably affecting the oil industry and clearly reflected in production, transportation, and price indexes. Included in these events are the Korean war (June 1950–July 1953); nationalization of the Anglo-Iranian Oil Company in Iran (March 1951); and Egypt's nationalization of

Table 4.1 **SOURCES OF ENERGY**
(Percentages)

	1913	1937	1950	1957
Oil	6	20	27	32
Coal	90	74	62	53
Hydroelectricity	4	1	2	2
Natural gas	...	5	10	13

the Suez Canal in 1956, which had been closed from October 1956 to March 1957.

NATIONAL FACTORS

Expansion in Venezuela's oil industry was evident in this period (1945–1958). The country's internal characteristics must be added to these international factors, because Venezuela already had a firm and developed oil industry, permitting it to benefit from favorable external events. Its production in 1946 of 1,064,326 barrels per day increased in subsequent years, exceeding 2,700,000 in 1957. In effect, between 1948 and 1957, Venezuela produced more than 7.6 billion barrels of crude.

Fiscal oil income exceeded $323 million in 1948 and had reached $595 million by 1955. In 1956 and 1957 it jumped to $970 million, but this was due to extra income derived from new concessions granted during those two years and representing $315 million and $370 million, respectively. During this period the percentage of participation of oil income in the national budget rose from 54 percent (1945) to 71 percent (1957).

Venezuela's national budget became more and more dependent on oil income. At the end of the three periods under review, percentage participation of oil income was: first period (1917–1936), 29 percent; second period (1936–1945), 54 percent; and third period (1945–1958), 71 percent. As the oil industry expanded, the country's economic dependency on it became more accentuated.

1946 Income Tax Law Amendement

In December 1945, Decree No. 112 was introduced, whereunder extra taxation was applied to incomes exceeding $259,000 originating from mining activities, oil among them. However, as this decree was limited to 1945, it was necessary to establish a norm that could be applied permanently to subsequent annual fiscal periods, for the same purpose of guaranteeing the nation a more equitable participation in oil-company profits. When the hydrocarbons law was debated and approved in 1943, it was decided that any unfavorable situation affecting the country's oil-profit participation could be eliminated by reforming the income tax law.

Herbert Hoover, Jr., and Arthur A. Curtis, United States experts who acted as consultants to the government in the 1943 hydrocarbons law, stated: " . . . application of the new Income Tax Law has not yet been incorporated in estimates of the oil industry's costs. As mentioned, this type of taxation is used as a form of control over excess profits obtained by individual and commercial companies. If used wisely, it could represent a further guarantee to the nation of its fair profit participation in industrial operations."[1]

The ideas expressed in that paragraph form one of the doctrinary foundations of the income tax law reform of 1946. In effect, it was considered recommendable to improve the nation's oil-profit participation by modifying the law, specifically by means of a complementary tax scale. This assumption is based on the report submitted by the Government Revolutionary Council to the National Constituent Assembly, in December 1946, stating: " . . . the government believes it is now time to reform (December 1946) covering "profits proceeding from noncommercial professions," "salaries, retirement, and other remunerations," and

Venezuela's Constituent Assembly decreed the income tax law partial reform (December 1946) covering "profits proceeding from noncommercial professions," "salaries, retirement, and other remunerations," and "complementary taxation." "Complementary taxation" concerns the oil industry. The existing tariff was entirely modified, all percentages being increased. In the 1944 law the highest percentage was 9½ percent, uniformly applicable from $647 million (Bs 2 million) up, whereas the 1946 reform applied this additional taxation from $259,000 (Bs 800,000), increasing it progressively up to 26 percent.

Through the complementary tax tariff modification, the nation obviously started receiving a higher oil-profit participation. However, it had still not reached its aim of at least equaling the oil companies' profits, and this situation persisted through subsequent years, during the enforcement of this tax reform.

The 1952 Memoir of the Ministry of Finance refers to agreements made between the state and Creole Petroleum Corporation and Compañía Shell de Venezuela to reach the parity goal through adjustments: "Hydrocarbon exploiting companies, including the Shell Group (Venezuelan Oil Concessions, Shell Caribbean Petroleum Company, etc.) and Creole Petroleum Corporation, have agreed by contract with the Government to pay a complementary amount to equalize the nation's participation in company profits of their 1946 and 1947 fiscal years. To that effect, it is agreed to retroactively apply provisions of Chapter XI, Additional Taxation, in the prevailing Income Tax Law, established for this purpose."[2]

These developments matured when Creole Petroleum Corporation paid under adjustment $4,782,606 (December 31, 1952) and the Shell group paid $1,432,770 (March 3, 1953). Payment of these adjustments

[1]*Exposition of Reasons,* Ministry of Finance, Caracas, 1968.
[2]Ministry of Finance Memoir, 1952, p. 350.

therefore shows that rulings of the years 1943 (hydrocarbons law), 1945 (Decree No. 112), and 1946 (reform of the income tax law) in practice failed to meet the nation's wish for an equal or higher participation in company profits. This situation remained unchanged until the 1948 reform referred to later.

Presidential Election—Rómulo Gallegos

On December 14, 1947, Rómulo Gallegos[3] was elected President of the Republic, taking office on February 14, 1948. Very special circumstances surrounding Gallegos's election made it a noteworthy event in the history of Venezuela's political institutions. This was the first time that Venezuelans had directly elected their President through popular, universal, and secret voting.

All twentieth-century Venezuelan constitutions prior to 1947 stipulated second- and third-degree indirect election systems for presidential elections and attributed electoral functions to various organizations.

Although nineteenth-century constitutions established presidential nomination through free elections, the system was not effective because voting rights were limited to citizens of certain age, sex, and economic and educational backgrounds, and hence, Venezuelans did not massively participate in these electoral procedures.

In a general sense it might be said that the election in 1947 was the first electoral process to take place in Venezuela with direct intervention by all sectors of the Venezuelan community capable of exercising this electoral right. Elections took place with participation by political parties, specific electoral organizations, all adults (including women) over eighteen years of age, and even the illiterate sector—i.e., within the widest concept of universal principles of suffrage.

The aspects relating to age, sex, and illiteracy had been contemplated in Venezuelan constitutional law to a greater or lesser degree of liberality and subject to being granted or rejected; however, in practice the people had never reached the voting booths to choose their President. The 1945 Constitutional Reform had granted voting rights to women, but only for municipal elections. The 1947 Constitution then granted all political rights.

Another prominent aspect lay in the implanting in the collective con-

[3]Rómulo Gallegos 1884–1969. After participating for many years in the teaching, literary, and political fields, he became the first President to be elected through popular vote. He is considered Venezuela's most representative novelist, not only because of the literary and social content of his books, but also because they are set in varied geographical regions of Venezuela.

sciousness the fact that the origin and source of political power lies in the popular vote, expressed through free elections. This concept has since become deeply ingrained in Venezuela's political formation, to the extent that when legal means have been sidestepped, the usurpers themselves have sensed a need to validate their spurious origin through popular vote, even though fully conscious that they would again deceive.

In Venezuela, where force, materialism, and arms had for a long time been predominant factors in the origin of governments, a new nonmaterialist ideological and moral phase of highly political consciousness began in December 1947: free elections. Although this first promising step fell into eclipse for ten years thereafter (1948–1958), it was the most significant political step taken by Venezuelans to that date.

The purpose of the December 1947 civic march was to directly elect the President of the Republic, congressional deputies and senators, deputies for state legislatures, and federal municipal council members. State municipal council members were elected subsequently, on May 9, 1948. Presidential electoral candidates were Rómulo Gallegos, for Acción Democrática; Dr. Rafael Caldera,[4] for the Social Christian party; and Dr. Gustavo Machado, for Venezuela's Communist party.

President Rómulo Gallegos (who had received an 80 percent majority) was overthrown after nine months in government (November 24, 1948). He had been elected for a five-year period that should have ended on February 14, 1953. He was ousted by a strictly military coup headed by top officers of the Armed Forces: Lieutenant Colonels Carlos Delgado Chalbaud, Minister of Defense; Marcos Pérez Jiménez a Chief of Staff, and Luis Felipe Llovera Páez, Assistant to the Chief of Staff. This was the climax to a series of subversive uprisings that had started after the October 1945 revolution, which deposed Gen. Isaías Medina Angarita.

By 1948 the political factors originating the October 1945 revolution had been overcome. In this three-year period the President of the Republic, members of Congress, and state and municipal legislators had been chosen by first-degree elections through popular, universal, and secret voting. The 1947 Constitution included these electoral reforms in its structure, and different sectors of public opinion gave birth to political parties, such as the Social Christian party,[5] which in time was to become powerful Acción Democrática's greatest rival.

The November 1948 military coup had no political justification or projection. It was a reactionary movement against changing events during

---

[4]Dr. Rafael Caldera was an unsuccessful candidate in the 1947, 1958, and 1963 presidential elections, but in 1968 he was elected President for the 1969–1974 constitutional period by a very small margin.

[5]Comité de Organización Politica Electoral Independiente, or COPEI.

those three years, and it initiated a dark stage in Venezuela's political institutions. Abuses committed from the very beginning confirmed the regime's reactionary and regressionist policy. Its negativism can be seen in its fundamental aim, which was to ignore popular will expressed through the 1946, 1947, and 1948 elections and to remove any person or group exercising power. Political parties and organized labor movements were declared illegal; Congress, other representative bodies, the municipal councils, and the Supreme Electoral Council were dissolved.

To mitigate this destruction of the country's institutional base, the military government's leaders offered no justification to the people, nor any outlets to the irregular situation arising thereby. Their explanations did not contain one single foundation, doctrine, or principle; they merely invoked the deposed government's administrative inefficiency (which in all due respect was quite true) and paradoxically offered eventual future elections: the return to the system against which they had risen.

I have referred briefly to these episodes in Venezuela's political life because of the consequences they were later to bring to the Venezuelan and international oil industry.

VENEZUELA AND THE
INTERNATIONAL OIL INDUSTRY

From October 1945 to November 1948, the Venezuelan government maintained and practiced measures which were a policy in themselves in facing the oil question. Taxes were increased with a view to increasing the nation's oil-profit participation, and to the purpose it was officially declared that no further hydrocarbon concessions would be granted. The first labor contract was drawn up in the oil industry, and a drive was started for establishing a national oil company. In summary, Venezuela was becoming conscientious in relation to the most important aspects of its oil industry.

President Gallegos, deposed in November 1948, in his message to the nation upon going into exile, inferred that oil companies had participated in his overthrow: "Powerful economic currents, those of Venezuelan capital lacking social sensitivity, and possibly those of foreigners exploiting Venezuela's oil wealth, not expected to favorably accept the limitations imposed upon them in fair defense of the people's welfare, by increasing their tax payments to the National Treasury, added to our decision not to grant further oil concessions, which are to become a source of reserve for the future of Venezuela, these—I emphatically repeat—are the facts behind the traditional lust for power which gave an impulse to this victorious military coup."[6]

The intervention by oil companies referred to by Rómulo Gallegos remains a point to be clarified by historians of this phase in Venezuela's political history.[7]

"Fifty-Fifty"—the 1948 Income Tax Law Amendment

On November 12, 1948, the income tax law was modified to include the new "additional taxation," applicable to taxpayers obtaining income from mining or hydrocarbons exploitation. The purpose of the new taxation was to split profits, on a fifty-fifty basis, between the taxpayer and the state; this reform became known in the oil and fiscal argot as the "fifty-fifty" reform.

In 1946 it was thought that the best way to assure 50 percent minimum state participation would be through increasing the complementary tax rate. However, the 1946 reform demonstrated that this was ineffectual, because when oil prices, costs of materials, and workers salaries varied, company profits also oscillated, affecting the equal participation sought by

[6]Historical documentation, José A. Catalá (ed.), Caracas, 1969.

[7]After President Rómulo Gallegos was overthrown, from 1948 to 1958 the country was governed first by a Military Council, up to the assassination of its presiding member, Lt. Col. Carlos Delgado M. Chalbaud, in 1950, and then by Lt. Col. Marcos Pérez Jiménez, who presided over a Government Board up to 1952 and then governed directly up to Jan. 23, 1958.

Pérez Jiménez governed by reinstituting a personalistic and coercive regime, using practices that were assumed to have disappeared after the death of Juan Vicente Gómez in 1935. Pérez Jiménez maintained a political police force—the National Security Guard—that unleashed a violent and cruel persecution against leaders and militants of political parties, especially Acción Democrática. Many were the citizens who were imprisoned, tortured, assassinated in prisons or in the streets, and sent into exile. He prohibited public freedom and disrespected human rights. He brutally assaulted the most varied sectors of the community; he silenced the press and closed down universities; he illegalized labor unions and political parties. Administrative corruption flourished to such a degree that Pérez Jiménez himself, after being overthrown, was extradicted from the United States in 1962, subjected to trial, and judged guilty of embezzlement by Venezuela's Supreme Court of Justice.

Those who supported and defended Pérez Jiménez's dictatorial regime attempted to justify it by representing it as a stage of progress of the country, citing the public works then built, which in a way did serve the subsequent progress of Venezuela. They also pointed to the fact that under his government, more order and personal safety prevailed in the country. Although these allegations are debatable, they do contain a certain degree of truth.

Using these arguments, Pérez Jiménez's partisans endeavored to restore him to government. However, the "perezjimenist" current, which in 1968 surprisingly obtained 400,000 votes and which in the five subsequent years structured the Nationalistic Civic Crusade party, was defeated in the 1973 elections, thereby proving that Venezuela's people are against dictatorial regimes and are determined to assure a democratic system in the country.

the nation. Therefore, in order to counteract this fluctuation, it was thought that the solution lay in creating an additional taxation on net profits earned by the companies, precisely where the desired principle of equality between the companies and the state could be applied.

Through this 1948 income tax law reform, it was the state's renewed intention to secure oil income which—added to the general and special taxes levied on the hydrocarbons industry—would at least be similar to the net profits received by producing companies. The additional tax amounted to 50 percent on excess net profits. This meant that normal taxes plus the excess tax would originate a profit relation of 50 percent for the taxpayer and 50 percent for the state.

As soon as this new taxation was enforced, taxpayers who were beneficiaries of oil royalties petitioned the Income Tax Board of Appeals and the Supreme Court of Justice to declare the new tax law unconstitutional. Numerous administrative and juridical procedures developed between 1949 and 1954 whereunder the nation supported the applicability of the additional taxation. Finally, the Court pronounced in favor of the nation in all these procedures, upholding the constitutionality of the taxation.

The Venezuelan Economic Mission

By private initiative through the Caracas Chamber of Commerce, a Venezuelan Economic Mission was organized for the purpose of informing the United States of the most important factors characterizing Venezuela's position in its commercial relations with the United States and also of the repercussions that measures restricting United States purchase of Venezuelan oil would have for both countries.

The mission visited the United States from March 9 to March 29, 1950, going to New York, Boston, Philadelphia, Pittsburgh, Detroit, Cleveland, Chicago, Milwaukee, St. Louis, New Orleans, and Washington, as these were the cities primarily involved in commercial relations with Venezuela. It directly contracted chambers of commerce, official bureaus, and business representatives of the industrial and banking spheres.

In Washington, the mission held meetings with top officers of the State Department (Messrs. Brown and Bauder), the Commerce Department (Messrs. Whitney and Blaidsdell), the Treasury Department (Mr. Martin), and the Interior Department (Mr. Swanson of the Gas and Oil Division). In all these interviews the mission stressed that it was strictly private and that its function was strictly commercial.

Subjects Treated by the Mission

The Venezuelan Economic Mission expressed the points of view held by the private commercial and industrial entities that it represented as members of the Caracas Chamber of Commerce. Such opinions were contained in the mission's agenda dated February 8, 1950. The mission also based its approach on an economic study prepared by the Central Bank of Venezuela at that time, relating to possible restrictive measures that were then the main topic in the United States. It presented a summary of Venezuela's position in 1950 concerning its foreign business relations with the United States.

It likewise pointed out the equality of treatment received by foreign investors in Venezuela, Venezuela's monetary stability, and the contradiction between United States commercial policy and its international agreements and President Harry S. Truman's declarations. In general, the points expressed and sustained by the Venezuelan Economic Mission (according to the report submitted to the private organizations it represented) can be summarized as follows:

1. That Venezuela was the United States' number one customer of all Latin American countries and second worldwide (excluding Great Britain and West Germany, which preceded Venezuela because of benefits obtained through the Marshall Plan).

2. That although the United States market absorbed 40 percent of Venezuela's total exports, Venezuela in turn bought from the United States 75 percent of its total imports, thus resulting in a commercial balance favorable to the United States.

3. That Venezuela maintained its business relations on a multilateral basis, the system proclaimed in international agreements and conferences as being the most appropriate for resolving the economic problems faced by most nations. That any United States restrictions on imports of Venezuelan oil would originate serious repercussions in the economic, political, and social life of Venezuela, which in turn would be unfavorable to the economy of the United States.

4. That Venezuela, because of its traditional policy of equal opportunity, was one of the most favorable countries for commercial exchange and investment of capital.

5. That if the United States restricted its oil purchases by creating custom barriers, it would contradict policy which it had been proclaiming as the fundamental basis of any international economic policy.

6. That Venezuela had unfailingly complied with commercial reciprocity treaties with the United States.

7. That Venezuela's traditional monetary stability and lack of import and exchange controls made it a highly favorable market for international trade.

8. That Venezuela had progressed without becoming a burden to North American taxpayers because, out of the loans granted by the Export-Import Bank of Washington, it had used only $27 million, repre-

senting 2.3 percent of the total granted to Latin America, and that it owed only $3,275,000 at the time.

9. That Venezuela had greatly increased its investments through its own resources.

Recommendations by the Mission

In its report, the 1950 Venezuelan Economic Mission made several recommendations, including:

a. That the favorable aspects of Venezuelan proposals should be further pursued in the United States.

b. " . . . the advisability and necessity to proceed wisely and actively in all spheres of national activity so that should oil exports be considerably restricted in the future, Venezuela would be in a better position to confront the situation with less risk and be more prepared to defend itself from an economic point of view."

c. " . . . that organizations representing industry and commerce should sponsor and undertake joint action to develop Venezuela's industry and *decrease the dangerous dependency* of its present economy."

d. " . . . that a Committee should be formed for the following purposes:

(1) To establish measures for expanding production.

(2) To ensure that existing industries complement their production processes by using Venezuelan raw materials.

(3) To concentrate efforts and available resources on agricultural and mining production.

(4) To create credit institutions for aiding in increasing that production considered most advantageous.

(5) To create permanent organizations representing private entities in order to coordinate production expansion."

These recommendations by Venezuela's private sector are valuable because of their logic and because they clearly describe the mission's objectives—defined by the matter of oil restrictions—of focusing on the most serious problem of the Venezuelan economy, that of Venezuela's economic dependency for the past fifty years on the oil industry. However, many difficulties must be overcome to put these recommendations into practice, and an even more serious obstacle is the persistent inaction by the government and by the private sectors. The recommendations have still not been put into effect.

The Ministry of Mines and Hydrocarbons

In Venezuela the matter of hydrocarbons was the specific function of the Ministry of Development up to 1950, when the present Ministry of Mines and Hydrocarbons was created. One department of this Ministry directly related to oil activities is the Hydrocarbons Technical Office, which is empowered by the state to control, inspect, and supervise the oil industry on a full scale, including preservation and maintenance of deposits and assets which will revert to the nation.

Hydrocarbon Concessions Granted in 1956 and 1957

Hydrocarbon concessions in force in 1955 covered 14,507,914 acres. In 1956 and 1957 the Venezuelan government granted new concessions, covering 2,028,952 acres, of which 769,147 were for exploration and 1,259,805 for exploitation. Combined, they originated an income of $684,776,318 to the National Treasury. Of the 1956 concessions, 276,048 acres located in Lake Maracaibo were for exploitation; another 469,498 acres located in the states of Zulia, Táchira, and Apure (bordering the republic of Colombia) were for exploration and exploitation.

Beneficiaries of the 1956 concessions were: Creole Petroleum Corporation, Compañía Shell de Venezuela Ltd., Mene Grande Oil Company, Venezuelan Sun Oil Company, Signal Exploration Company, The Superior Oil Co. of Venezuela, Venezuelan American Independent Oil Producers Association Inc., and San Jacinto Venezolana. By December 31, 1956, hydrocarbon concessions in force covered 15,250,960 acres.

During 1957, concessions granted totaled 1,280,930 acres, i.e.: 791,599 acres for exploration and 489,331 for exploitation. The grant of these concessions represented an income of $368,987,402 to the National Treasury. These concessions were located in Lake Maracaibo, in the states of Barinas and Monagas, in the Gulf of Paria, and in zones bordering the republic of Colombia.

Beneficiaries of these concessions (the last granted in Venezuela) were the following: Compañía Shell de Venezuela, Creole Petroleum Corp., Venezuelan Sun Oil Company, Venezuelan American Independent, Continental Oil Company of Venezuela, Phillips Petroleum Co., Venezuelan Atlantic Refining Co., Pan Venezuelan Oil Co., San Jacinto Venezolana C.A., and King-Mill Oil Co. C.A.

No hydrocarbon concessions had been granted in Venezuela since 1945 under the government of Gen. Isaías Medina Angarita. In other words, from 1945 to 1956 the official policy prevailing in the country had been not to grant any further concessions. The interruption in 1956 of this policy—which had been upheld throughout the years—was because the country faced the fact that in order to expand hydrocarbon reserves, it would be necessary to strengthen the oil industry and increase the relationship between territory granted and barrels produced. One of the principles of these new concessions was that of geographical priority in order to accelerate exploitation of areas adjacent to Venezuelan frontiers and other regions recommendable on account of technical and economic factors. Another influential reason for granting concessions was the financial difficulties confronting the government at that time.

I shall at this point conclude the first part of this book, covering the development of Venezuela's oil industry. After 1958 many negative international events took place which reflected on Venezuela's oil industry and which influenced the new political, economic, and social developments that in later years markedly characterized the country, as we shall see in the following chapters.

The Present—
Decline

I am concerned about the present and also about the future.

SIMÓN BOLÍVAR
Letter to General Robert Wilson,
Bogotá, November 13, 1827.

FIVE

The Decline:
Contributing International Factors

I shall now analyze the decline of Venezuela's oil industry, which started in 1958 and has continued to the present. In characterizing this present stage as one of decline, I refer to the industry proper, i.e., to the decrease in its reserves and production capacity; to the situation of its assets; to the diminished investments, etc. The general change in prices and the state's profits per barrel are not taken into consideration, because these aspects are not a precise and direct gauge of the growth or decline of the industry proper.

BACKGROUND

Several negative international factors contributed to the decline of the productive capacity of Venezuela's oil industry. Many of these factors accumulated during the preceding phase of expansion, only to crystalize after 1958. These international factors affected not only Venezuela but also other exporting countries. During the 1960s the international situation changed to a point where at the end of the decade an era of international oil boom had developed. This era was marked by a pronounced expansion of markets and greater power by exporting countries, which in general entered a stage of prosperity through the expansion of oil-industry installations. The notable exception was Venezuela.

Venezuela was the country that initially promoted the founding of the Organization of Petroleum Exporting Countries (OPEC) in 1960, and although it has greatly profited from the market and price booms of the 1970s, the decline of its oil industry proper, which started at the end of the 1950s, has continued. As a consequence of internal factors that I shall also analyze, Venezuela's industry has been subjected to a process of decapitalization and lack of investment that has caused a gradual decrease in its proven reserves and hence a decrease in its productive capacity.

It is therefore a paradox that, at the beginning of this era of expansion and profits, without precedent in the history of oil-exporting countries, Venezuela—which greatly contributed to the boom through the creation of OPEC and in pointing out new ways to this group of countries—is not in a position to take full advantage of the situation because of its limited productive capacity. Furthermore, Venezuela's oil industry is declining to such an extent that, over a short span, its gradual production shrinkage could deteriorate its position within this continuously more powerful group of oil-exporting countries.

Middle East Oil Expansion

The negative factors which accumulated during the expansion stage of Venezuela's oil industry (1945–1958) have contributed to its decline. First of these is the great oil expansion in the Middle East. As mentioned previously, from 1926 to 1947 Venezuela's production exceeded that of all Middle East countries. After 1947, it was surpassed by the joint production of Iran, Saudi Arabia, Iraq, Kuwait, and Bahrain, whose costs were far below those of Venezuela and who already had by then nearly one-third of the world's proven reserves. By 1958 Middle East production exceeded 4 million barrels per day, compared to Venezuela's 2.5 million, and the gap has become wider as the years have gone by. It must be remembered that prior to the 1960s the oil market was dominated by buyers and not by sellers as it is at the present. Therefore, the production growth in the Middle East—Venezuela's main competitor—has negatively affected the development of Venezuela's oil industry.

Table 5-1 shows Middle East and Venezuelan production expansion from 1945 to 1972. It can be seen that while production doubled in the Middle East between 1965 and 1972, in Venezuela it diminished by 250,000 barrels per day. In 1950 the return per well in the Middle East was eighteen times that of Venezuelan wells, implying much lower production costs; furthermore, labor costs and social benefits in the Middle East were then—and still are—much lower than those prevailing in Venezuela.

Nevertheless, Venezuela had the advantages of its geographical location and its greater development and maturity in the various phases of its oil industry, which enabled it to maintain its position in the United States

Table 5.1 **PRODUCTION OF CRUDE, 1945–1972**
(Thousands of barrels per day)

Year	Middle East	Venezuela	Year	Middle East	Venezuela
1945	512	886	1959	4,560	2,771
1946	680	1,046	1960	5,200	2,846
1947	813	1,191	1961	5,576	2,920
1948	1,109	1,339	1962	6,110	3,200
1949	1,371	1,321	1963	6,701	3,248
1950	1,725	1,498	1964	7,358	3,393
1951	1,889	1,705	1965	7,995	3,473
1952	2,048	1,804	1966	8,877	3,371
1953	2,393	1,766	1967	9,454	3,542
1954	2,706	1,895	1968	10,358	3,605
1955	3,215	2,157	1969	11,268	3,594
1956	3,408	2,457	1970	12,309	3,708
1957	3,504	2,779	1971	15,041	3,549
1958	4,221	2,605	1972	15,400	3,219

SOURCE: Ministry of Mines and Hydrocarbons, Caracas.

market and, for a time, in the European market. However, since 1959 Venezuela's oil has gradually been displaced by Middle Eastern oil in European as well as Latin American markets. Venezuela's competitive status slipped further, even in the United States, due to the larger tankers built as a consequence of the first Suez crisis.

African Production

The second negative factor is the development of production in Africa. During the expansion that characterized the preceding stage (1945–1958), African oil had not yet reached the market, because its industry was still in the process of exploration and installation. While it is hard to pinpoint the timing and the degree of influence that the Middle East's extraordinary oil expansion has had in relation to the decline of Venezuela's oil industry, it is possible to do so in relation to the arrival of African oil on the market—from Libya, Nigeria, and Algeria. The combined production of these three countries, up to 1958, was practically nil (87,000 barrels per day in 1958). However, the higher quality of African crudes (Venezuela's historical average is within the range of 25° API[1] and Africa's is close to 30° API), the closeness of this region to the large Western European consumer centers, and the increased volume of its production

[1]American Petroleum Institute gravity.

after 1959 were important factors in displacing Venezuelan oil from those markets and in generating a surplus which, as we shall see later, caused a progressive drainage on the general level of oil prices.

It is a known fact that any industrial process—be it new or expansional—generally requires a certain length of time to mature, sometimes many years. This is particularly so in the petroleum industry. The installation of Africa's industry, one of the most important to be incorporated in the last forty years, was accelerated after the Suez crisis in 1957; however, exploration and development activities required many years. As we shall see in Chapter 10, these facts should be taken into particular account by Venezuela. If Venezuela wishes to reactivate its oil industry, it must add, by means of intensive exploration and exploitation, new petroleum deposits to this vital economic activity, at present in a state of decline which many Venezuelans (especially the country's leaders) refuse to admit.

The Supertankers

A third factor negatively influential to Venezuela lies in the constant growth of an international fleet of tankers to which larger and larger units are continuously incorporated. The first Suez crisis in 1957 dramatically pointed out the risk international companies were running in relying so pronouncedly on this route of communication, the Suez Canal, for transporting oil from the Middle East. There was, of course, the alternative of going around Africa through the Cape of Good Hope, but this meant a substantial increase in tanker freights, the deadweight of which in most cases was less than 25,000 tons (close to 80 percent of the world's fleet in 1956 was composed of tankers of less than 25,000 tons).

This circumstance therefore speeded up plans for building larger tankers, which began transporting oil from 1958 in increasingly greater proportions. The second Suez crisis, which occurred ten years later, become another factor that precipitated the construction of giant tankers. These larger tankers have a deadweight of close to 500,000 tons and in a relatively short time could reach one million tons.

Consequently, by the late 1950s Venezuela had to confront this new development, the supertankers, which contributed to defeating one of Venezuela's most competitive advantages: its geographical location. This competitive factor, negative to Venezuela's oil in consumer markets, was engendered during the period of Venezuela's industrial expansion, but its effects were to be felt later. The greater the tanker capacity, the lower the cost per barrel transported, which served to reduce Venezuela's competitive advantage in this area. At present the cost per unit transported is cheaper via giant tankers than via traditional tankers, although the latter are still being built for transporting refined products and for carrying crude from ports such as Venezuela's that lack adequate installations for ships of great draught. Large tankers can be loaded and unloaded faster than smaller tankers, their fuel consumption is not proportional to the increased loading capacity, and they do not require a larger crew. (Approximately 30 men are sufficient either for a tanker of 25,000 tons or for one of 150,000 tons.)

Posted Prices

A fourth negative factor is the shift in "posted prices,"[2] which were traditionally based on Gulf of Mexico prices. In my opinion, there were three fundamental reasons for this shift in prices quoted for petroleum produced outside the United States, causing their decrease: the great production expansion in the Middle East; the development of production in Africa; and the fact that the United States became a net oil importer in 1948.

The United States, prior to 1948, was a petroleum exporter, interested in maintaining high prices for crude and refined products. After 1948 it had to buy these products in increasing quantities from foreign producing centers, especially those in the Caribbean zone (Venezuela). To Gulf of Mexico prices (FOB East and West Texas prices), on which prices of crude oil produced outside the United States were based, transportation costs to New York had to be added. Gulf of Mexico FOB prices were very close to those quoted in Venezuelan ports, because the distance separating the two ports from New York was almost the same.

Before 1950 the point of equilibrium for world quotations was Naples, Italy. In those days, prices of Middle East crude were based on Gulf of Mexico prices, plus the cost of transportation to Naples, less freight between Naples and the Persian Gulf. The same procedure applied to Venezuela, i.e., to Gulf of Mexico FOB prices were added freights to Naples, and from the resulting price the cost of transportation from Venezuela to Naples was substracted.

Temporarily, the point of equilibrium was transferred to London. Finally, about 1950—two years after the United States became a net oil importer—New York's great market was taken as a point of reference. Posted prices started appearing in the following areas: for petroleum by-products, the Caribbean zone in 1950, Singapore in 1955, and the Persian Gulf in 1957; for crudes, the Persian Gulf in 1950, Venezuela in 1952, Sarawah (Malaysia) in 1954, and North Africa in 1958.

---

[2]"Posted prices" are FOB prices port of origin of crudes, and they represent the value at which companies are willing to sell their production.

It is important to bear in mind that prices change from one country to another and even vary within the same country, depending on the source of supply. Therefore a wide range of circumstances influences pricing in any given region.

Oil prices began declining after 1959. Up to 1961 Iraq suffered most, with a decrease of 18 cents per barrel; then Qatar, 15 cents, followed by Saudi Arabia, Iran, and Kuwait, with decreases of 13 cents per barrel.

Refining

A fifth factor that was to influence the decline of Venezuela's oil industry—like the others, engendered during the preceding phase of expansion—lies in the development of refining within the great oil-consuming centers. Before the Second World War almost 70 percent of crude was refined within the producing contries, which consequently exercised certain influence in product pricing in consuming areas. But after the war ended and toward the beginning of the 1950s, with the great economic expansion of the United States, Japan, and European countries, a refining capacity was established in these nations which, in some cases, exceeded their own needs. This development meant that oil-producing countries could influence pricing only in those consuming nations that were unable to cover their own needs for refined products. The developed nations, great consumers of raw material for their industries, thereby created conditions disadvantageous to underdeveloped countries, suppliers of such raw materials. There were, of course, strategic, economic, and technical reasons that influenced this policy.

Among the strategic reasons was the need for industrial nations to have their own refineries so as not to depend on by-products obtained at export centers, which were subject to risks of war or confiscation. For instance, the Abadan refinery in Iran was confiscated under the government of Prime Minister Mossadegh.

Economically, massive transportation of crude to great consuming centers by supertankers is cheaper than carrying naphtha, gasoline, gas oil, kerosene, etc., in small tankers or in compartments of supertankers. Furthermore, it is more practical and economical to load and unload great volumes of crude than relatively small quantities of by-products. It is likewise important to keep in mind that in the postwar years European countries sought, as one of the principal goals in their economies, the currency saving that refining in their own territories represented.

While this was happening at an international level, Venezuela's policy of increasing the value of its petroleum by means of refining in Venezuela, as proclaimed in the 1943 hydrocarbons law, began having a large-scale effect. At the beginning of the 1950s its great refineries at Amuay and Punta

Cardón started operations. Thus, development of refining in Vene-zuela—the fruit of a very logical national policy—coincided with a similar process in the great consuming centers, principally the United States and Europe, Venezuela's major oil markets.

It is my opinion, however, that this situation exercised a great influence in the change that took place in the composition of Venezuelan petroleum sales, bringing negative results to the country. I refer to the ever-increasing volumes of fuel oil in Venezuelan exports. In 1947, Venezuela exported nearly all its crude (94 percent) and only 6 percent in refined products. In 1950 crude exports dropped to 87 percent, fuel oil increased to 9 percent, and light products made up 4 percent. In 1958 this relation-ship became 77 percent, 15 percent, and 8 percent, respectively, and in 1972 it became 69 percent, 24 percent, and 8 percent. It can therefore be seen how the proportion of fuel oil, produced and exported, increased.

It is significant to bear in mind yields in refining throughout the world and in Venezuela. In 1950 fuel oil production in world refining repre-sented 32 percent of the total, and in 1966, 27 percent. But in Venezuela the proportion was 57 percent in 1950 and 60 percent in 1966. It is true that the heavier weight of Venezuelan crudes implied a higher yield in heavier by-products; however, these same crudes rendered higher yields of light products when processed at refineries in industrialized nations, such being the purpose of those refineries. Since the price of fuel oil was less than the price of crude, notwithstanding the added value of the refinery process, it is therefore regrettable that the Venezuelan refineries were yielding more than twice the world average for fuel oil.

This exaggerated participation in the production of fuel oil in Venezue-lan refineries is more or less similar to that of the Aruba and Curaçao refineries, which in relation to the international markets might be consid-ered an extension of Venezuelan refining (at least until recently). This is a direct consequence of the refining policy followed by industrial nations, especially that of Venezuela's principal market, the United States.

By examining the refining situation in Latin America, we can see a similar tendency toward increased activity within each country's own territory. Refining capacity in Mexico, Central America, and South Amer-ica (excluding Venezuela) by 1957 amounted to 664,000 barrels per day in comparison with a consumption of 920,000 barrels per day. It can thus be seen how markets in this area were also being closed to Venezuelan refined products.

The Suez Crisis

The sixth negative factor is the Suez crisis itself. In 1957 it precipitated a drastic increase in Venezuelan production, but when it subsided in 1958, the country's production and exports to international markets decreased considerably, thus contributing to a surplus in the following years.

Independent Producers

A seventh important factor was the entry into the world market of the so-called independent producers. Up to the end of the 1950s integrated oil companies—the seven big companies—maintained an almost total monopoly over all international oil operations. It is estimated that in 1959 these great companies controlled (excluding the United States and the Communist countries) 85 percent of world production, nearly 75 percent of refining, and close to 70 percent of trade.

In the mid-1950s a new group of United States oil companies, which up to then had operated principally in the United States, entered the international field, obtaining concessions in Venezuela, the Middle East, and Africa, their main purpose being to secure oil supplies for their plants in the United States. These companies became interested in international operations because of increased world demand for oil during the postwar era, offering a possibility to increase sales within the United States and at the same time capture new markets abroad. Furthermore, the Suez crisis had originated an increase in oil prices, thus expanding the promising panorama the new international companies visualized.

However, as these independents were not fully integrated, some of the aforementioned factors placed them in a critical position. The return to normal price levels, the reduction in freights, and the decrease in the rate of growth of world demand caused an excess of supply. Furthermore, although it was initially thought the independents' production would be absorbed, the United States established restrictions on its oil imports (voluntary and later mandatory), thus closing the doors to those companies trying to enter the market with their own production.

The new companies—or independents, as they were called to distinguish them from the great integrated groups—had spent considerable amounts of money in obtaining concessions abroad and had made heavy investments that *had* to be recuperated. Production proceeding from these companies aggravated the surplus situation and encouraged an even further decrease in prices, as the independents began granting special discounts in order to secure markets for their production.

Russian Oil

A further negative factor was the entry, toward the end of the 1950s, of Russian oil in world markets. Russia became relatively important as an oil exporter: by 1960 its exported volume amounted to 635,000 barrels per day, representing 20 percent of its production. Russian oil sales to the rest of the world, like its trade in general, responded to both political and economic reasons. Politically, Russia's interest in extending its influence in developing countries led to offers of technical and economic assistance under conditions advantageous to several Third World countries. An important economic reason was its interest in importing products derived from Western technology, for which it required dollars or European currencies.

Price Deterioration

The oil restrictions established by the United States discriminated against Venezuelan oil in relation to Mexico's and, most especially, Canada's, and contributed to the industry's international crisis toward the end of the 1950s and beginning of the 1960s. In view of the significance of this event, we shall examine certain of its most important aspects, such as the resultant price slump that whipped industry exports (except those of the highly protected United States).

Even though posted prices do not reflect the true level at which oil sales are closed, they are a good index of the fluctuations that occurred in oil

Table 5.2   **POSTED PRICES, 1950–1959**
(Dollars per barrel)

Year	Algeria[a]	Iran[b]	Iraq[c]	Kuwait[d]	Saudi Arabia[e]	Vene- zuela[f]	United States[g]
1950	. . .	. . .	. . .	. . .	1.75	2.57	2.68
1951	. . .	. . .	1.67	. . .	1.75	2.57	2.68
1952	. . .	. . .	1.67	. . .	1.75	2.57	2.68
1953	. . .	. . .	1.92	1.72	1.97	2.80	2.92
1954	. . .	1.91	1.92	1.72	1.97	2.80	2.93
1955	. . .	1.91	1.92	1.72	1.97	2.80	2.93
1956	. . .	1.91	1.87	1.72	1.97	2.80	2.93
1957	. . .	2.04	1.98	1.85	2.08	3.05	3.30
1958	2.90	2.04	1.98	1.85	2.08	3.05	3.20
1959	2.77	1.86	1.80	1.67	1.90	3.05	3.20

[a]40° API and over, ex Bougie.
[b]34° API–34.9° API, ext Kharg Island.
[c]35° API–35.9° API, ex Fao.
[d]31° API–31.9° API, ex Mena.
[e]36° API–36.9° API, 1950–1956.
34° API–34.9° API, 1957–1959, ex Ras Tanura.
[f]35° API–35.9° API, ex Puerto La Cruz.
[g]30° API–30.9° API, ex Gulf of Mexico.
SOURCE: *20th Century Petroleum Statistics*, Degolyer and McNaughton, Dallas, Texas, 1971.

markets during the 1950s, particularly in the United States market, because the point of reference for price setting was the port of New York. We have seen that to Gulf of Mexico (Texas oil) value of crudes (at the well), freight costs to New York are added; from this resulting price, all costs applicable between the port of origin abroad and the port of New York are deducted. The result is the posted price, or FOB price of oil-exporting countries, at port of shipment.

Such a system of calculation divorced the price of crude in the producing areas from its true value; it is easy to see that production costs within the producing countries did not influence pricing. Nevertheless, posted prices proved to be good indexes of oil sales and served as a comparative basis for negotiation between buyers and sellers. Table 5-2 shows the general price situation in the decade of the 1950s.

It can be seen that these prices were quite stable, with only slightly rising tendencies. The upset occurred toward the end of the decade as a consequence of the first Suez crisis, which marked the series of events described in this book that led to the creation of OPEC in 1960. In effect, the Suez crisis caused an increase in oil prices during 1956 and 1957, but thereafter, once the Suez emergency had subsided, prices fell below levels prevailing before the crisis (1950–1952). Table 5.3 reflects price variations from the Suez crisis to 1961:

Table 5.3 **PRICE VARIATIONS, 1956–1961**
(Cents per barrel)

Year	United States	Vene-zuela	Saudi Arabia	Iraq	Iran	Kuwait	Qatar
1956–1957	25	24	9	1	7	9	7
1957–1958	0	1	6	7	6	4	6
1958–1959	−19	−22	−17	−16	−16	−16	−16
1959–1960	−2	−3	−5	−5	−6	−5	−5
1960–1961	0	0	−6	−5	−4	−5	−7
Net variation during period	4	0	−13	−18	−13	−13	−15

Note: The minus symbol indicates decrease.

SOURCE: *20th Century Petroleum Statistics,* Degolyer and McNaughton, Dallas, Texas, 1971.

It is to be noted that the United States posted price, at the end of 1961, had increased by 4 cents, and Venezuela's had returned to its old level, whereas Middle East crudes had decreased by 13 to 18 cents per barrel. After transit was restored through the Suez Canal, and the relative oil shortage in the world markets had been overcome, prices tended to resume their previous levels. The United States took the initiative by reducing 19 cents per barrel, but the Middle East followed this first readjustment by others, bringing prices far below their prior levels.

The British Petroleum Company was attributed to having started the 1959 decrease in crude prices.[3] It was immediately followed by other companies, to the detriment of export countries whose oil income was based on posted and closing prices. The closing price is the real sales price established through a variable discount on the posted price, and therefore modifications to the latter affect real prices and incomes of producing countries.

The price deterioration prompted Venezuela to create, in 1959, the Coordinating Committee for the Conservation and Commerce of Hydrocarbons (CCCCH), an organization responsible for controlling Venezuelan oil prices and empowered to prohibit negotiations that could adversely affect the country's interests. At the international level, the Organization of Petroleum Exporting Countries (OPEC) was created in 1960 for the fundamental purpose of "stabilizing international market prices." Its creation and development are described in a later section of this book.

Once OPEC was established, its members drew up a set of rules tending to carry prices to their 1956 level. In the beginning hesitantly and later vigorously, reference prices were imposed. These reference prices, based on values assigned to crudes and unrelated to posted prices, enable OPEC members to obtain a higher fiscal participation, since these values represent levels at which each country wishes to sell its production, even though in practice such levels are not always reached.

OIL RESTRICTIONS

It is well known that Venezuela's main oil market is the United States of America. More than 40 percent of Venezuelan exports are destined to the four main oil districts in the United States, especially those in the East. During the Second World War, the Petroleum Administration for War divided the United States into five oil districts: District I (East Coast), District II (Mid-Continent), District III (Gulf Coast), District IV (Rocky Mountains), and District V (West Coast). Ninety-five percent of Venezuela's oil exported to the United States was channeled to the first four districts. District V was Canada's natural oil market.

Table 5.4 shows Venezuelan exports of crude oil and by-products to the United States from 1954 to 1958. Although the United States is the

[3]At that time the Venezuelan government sent a complaint to the British government over the British company's action.

Table 5.4   **VENEZUELAN EXPORTS TO THE UNITED STATES, 1954–1958**

(Barrels per day)

Year	Crude	By-products	Totals
1954	356,733	350,387	707,120
1955	388,207	430,315	818,522
1956	450,533	472,158	922,691
1957	538,526	531,803	1,070,329
1958	441,519	606,121	1,047,640

world's first oil-producing country,[4] it is also the foremost importer, since its production does not entirely cover its own demand. Because of this, Venezuela's petroleum exports to the United States since 1947 shows a constantly increasing tendency for covering United States internal deficits, and amounted to approximately 12 percent in 1958.[5]

By 1957, oil imported by the United States proceeded from the following countries:

Venezuela	53.3%
Canada	15.7%
Kuwait	12.8%
Saudi Arabia	5.0%
Colombia	2.2%
Mexico	0.7%
Others	10.3%

From these figures it is obvious that, while the United States is Venezuela's most important oil market, Venezuela in turn represents the United States' foremost source of supply. During three consecutive decades, and since Venezuela displaced Mexico in 1928 as the United States' main supplier, Venezuela has occupied first place among the twenty-seven countries supplying the United States,[6] a situation which, although to a lesser degree, has continued to the present.

By "oil restrictions" is generally meant the limitations place on oil imports to the internal market of the United States through official measures applied by the United States government: custom tariffs and quota systems; voluntary or mandatory import programs; and quality controls.

Up to 1932 oil was imported freely by the United States, but in that year

[4]In 1957 United States production reached 6,459,300 barrels per day, which represented 41 percent of world production. Its imports covered 17 percent of its internal consumption.

[5]During 1973 the daily average of crude and by-products amounted to 6.2 million barrels per day, i.e., 35 percent of its total requirements.

[6]In 1928, Venezuela exported 21,987,000 barrels to the United States.

the Revenue Act taxing oil imports was approved. Although this legal ruling fulfilled objectives in the tax field, it constituted restrictive measures in relation to oil imports. United States imports of oil from Venezuela were then governed by successive reciprocal trade treaties between the two countries from 1939 to 1972, when they were no longer applied.

Voluntary Import Programs

A recurrent campaign in the United States against oil imports, begun in 1954 and supported by the so-called independent producers—among them the Texas Independent Producers and Royalty Owners (TIPRO) and the Independent Producers American Association (IPAA)—ended when the United States government established voluntary import programs and mandatory import programs in 1957–1958, commonly known in the oil world as the oil restrictions.

The first-mentioned were classified as voluntary, because their provisions were not obligatory, but rather indicated the objectives desired for maintaining an equilibrium between the United States oil industry and badly needed imports. On the other hand, as their name so indicates, the second were mandatory.

The voluntary programs established by the United States government were as follows:

Voluntary Program Dated July 29, 1957

Its objective was to maintain oil imports at the level of the average figure corresponding to the years 1954, 1955, and 1956, less 10 percent. The restrictions were applied to the first four United States oil districts, i.e., the most important in the country. The fifth district was subject only to restrictions covering crudes, and not derivatives. Consequently, oil destined for the fifth district was not heavily affected by this first restrictive program.

Voluntary Program Dated January 1, 1958

This program differed from the preceding one in several respects; for example, it was applicable to all five United States oil districts without exception, but with variations provided for each of them.

Voluntary Program Dated June 3, 1958

In this program, import limitations were extended to semiprocessed gasoline and to other semifinished products, taking as a basis the daily average of the first five months in 1958.

In principle, these programs were not discriminatory in the sense of establishing quotas for each exporting country, but in practice, they were discriminatory, because they were not applicable to imports via land, so they granted preferential treatment to oil proceeding from Mexico and Canada.

Mandatory Restrictions

The voluntary restrictions failed to curb the growth of United States imports of crude oil and by-products, and this failure was partly attributed to the expansion of the independents, the companies that had recently entered the world market with their own petroleum sources. Between 1954 and 1958, the number of independent United States oil importers had increased from 11 to 55.

United States oil production was increasing at a 1.8 percent annual average, whereas internal demand duplicated that rate. This imbalance contributed to the failure of the voluntary import programs. With demand increasing more rapidly than production, a breach was opening which had to be covered by means of additional imports of crude and by-products. Oil imports reached 1,700,000 barrels per day in 1958, nearly 700,000 barrels more than in 1954, for an average increase of 12.7 percent per annum, as shown in Table 5.5.

Between 1957 and 1958 United States internal demand increased by 3 percent, whereas production decreased by 4.9 percent, as can be seen from Table 5.5. Therefore, total imports increased by 8 percent. (We can see how in actual fact the increase in imports occurred in relation to finished products and not in relation to crude, which decreased by 6.8 percent.)

This unusual situation apparently alarmed President Eisenhower's gov-

Table 5.5   **U.S. OIL SUPPLY AND DEMAND, 1954–1958**
(Thousands of barrels per day and percentages)

	1954	1955	1956	1957	1958
Demand*	8,115	8,827	9,209	9,386	9,358
Production†	6,342	6,807	7,151	7,170	6,819
Importation:					
Crudes	656	782	934	1,022	953
Products	396	466	502	552	747
Total	1,052	1,248	1,436	1,574	1,700
Imp./demand, %	13.0	14.1	15.6	16.8	18.2
Imp./produc., %	16.7	18.3	20.1	22.0	25.0

*Includes exportations.
†Does not include hydrocarbons different from crude oils.
SOURCE: *20th Century Petroleum Statistics,* Degolyer and McNaughton, Dallas, Texas, 1971.

ernment, which, on March 11, 1959, introduced the so-called Mandatory Oil Restrictions program, applicable to imports of crude and nonfinished products. A little later, on April 1 of that same year, a second program regulating imports of finished products and fuel oil was announced.

These two dates mark the beginning of what Venezuelans know as the discriminatory treatment of their oil. It is to be noted that in introducing such measures, President Eisenhower relied on the Trade Agreements Act, which authorizes the President of the United States to restrict excessive imports of certain products should they become a menace to national security. In actual fact, the commercial treaty governing relations between Venezuela and the United States did contain a clause (Article XIII-Bis) to this effect.

The Venezuelan government, perceiving the barriers that such mandatory restrictions raised for its oil, addressed the United States government requesting treatment equal to that granted Canadian oil. The mandatory program tacitly excluded Canadian oil from the restrictions, because hydrocarbons imported via land to the United States were not subject to control.

I have so far omitted any reference to Mexico. Oil from Mexico entering the United States followed a *sui generis* scheme, known as the Brownsville Loop. Exclusion of imports via land from the restrictions program was not directly applicable to Mexico, because no pipeline existed uniting it with its North American neighbor. Considering the small volume of oil involved, it would have been far too costly to lay a pipeline to take advantage of the free access to the United States market. Nor was there any point in thinking about using trucks or tanker trucks, because of the high costs involved in that type of oil transportation.

The problem was solved in a very shrewd and unorthodox way. Protected by norms regulating the entrance into the United States of merchandise in transit to other countries, the Brownsville Loop started operating: oil despatched in tankers *from* Mexican ports entered Brownsville (a Texas city at the Mexican border) in transit *to* Mexico, where it was sent by land; it was then reexported to the United States again by land, thereby formally complying with the clause excluding such imports from the restrictions.[7]

The Venezuelan government's concern about dangers implied in the

---

[7]This situation was then legalized, eliminating the aforementioned loop and allowing imports of Mexican oil without specifying its route of entry to the United States, through President Nixon's declaration dated Dec. 22, 1970, in which he stated that "after Dec. 31, 1970, provisions controlling imports of crude, semifinished, and refined products would not be applicable to Mexico. . . . Mexican oil and its derivatives could enter into Districts I–IV and V without requiring quotas or licenses . . . such imports not to exceed an average of 30,000 barrels per day in each calendar year."

mandatory restrictions came to light shortly thereafter. The first program stipulated that imports, excluding those from Canada and Mexico, were not to exceed 9 percent of the United States's internal demand; this percentage was later increased to 12.2 percent, but it included all imports. In other words, oil imports proceeding from the rest of the world, principally from Venezuela, were determined after deducting the amount of oil proceeding from Canada and Mexico from the 12.2 percent quota, thereby impairing imports from other areas.

Throughout the period in which the mandatory restrictions program was enforced (1959 to 1970), demand in the United States market maintained an average annual increase of 3.8 percent, and internal production an increase of 2.9 percent. In other words, the ratio of internal supply to demand continued at a disequilibrium. Consequently, imports increased at an annual rate of 6.1 percent during this period.

I shall now analyze United States import figures in fuller detail:

1. In 1970 United States imports of crude and by-products represented 23.4 percent of demand and approximately 35.5 percent of internal production.

2. Imports of crude increased from 965,000 barrels per day in 1959 to 1,324,000 barrels per day in 1970, for an average annual increase on the order of 2.9 percent.

3. Imports of by-products increased from 814,000 barrels per day to 2,094,000 barrels per day between 1958 and 1970, with an equivalent average annual increase of 9 percent. However, fuel oil represented nearly three-fourths of total by-products.

4. The United States' main fuel-oil supplier is Venezuela; nearly three-fourths of that consumed in the United States proceeds from Venezuela. In turn, fuel oil sold by Venezuela to the United States represents approximately three-fourths of the total products exported to that country, and more than half of the total exports.

5. Increased imports of fuel oil are due to a well-defined policy adopted by United States refiners (and also by those of other developed nations), which considered that production of this petroleum derivative was not too attractive at the price levels established in their markets.[8] Therefore, refiners started decreasing their production capacity of fuel oil to a point where they could, as a result, increase that of other, higher-priced by-products.

Internal production of fuel oil decreased from 1,286,000 barrels per day in 1951 to approximately 953,000 barrels per day in 1959, the year in which the system of mandatory import restrictions was introduced. This represented an average annual decrease of 3.7 percent, whereas internal

---

[8]Prices of fuel oil nearly always have been lower than those of nonprocessed crudes.

demand decreased by only 0.02 percent during the same period, from 1,546,000 barrels per day in 1951 to approximately 1,543,000 barrels per day in 1959. The breach between internal United States production of fuel oil and demand has contined to widen, because production during 1970 covered only 32 percent of demand, against 83 percent in 1951 and 62 percent in 1959.

On the other hand, the thesis that Venezuela suffered discriminatory treatment is supported by the following data: Between 1959 and 1970, United States crude-oil imports increased by more than 350,000 barrels per day, which is equivalent to a 2.9 percent annual increase. United States imports from Canada increased at a 19.8 percent annual average, whereas crude proceeding from Venezuela decreased by an average rate of 4.7 percent. Displacement of Venezuelan crude in the United States market is obvious.

Two interesting facts in relation to Canada are:

1. The expansive rhythm of its crude exports to the United States is basically limited only by Canada's production capacity. Therefore it was also logical that the United States should be interested in increasing the rate of exploration in Canada, in order to incorporate new reserves and to maintain, or increase, the production capacity of its northern neighbor.

2. Canadian imports of Venezuelan crude have also increased. Even so, these imports are relatively small, in comparison to Canada's sales to the United States.

It seems strange that a country which has surplus oil for export should have to import it, but if one is aware of Canada's geography, it is easier to understand. The main oil fields are located in the West, whereas the great consuming areas are located in the center and East. Consequently, taking into consideration the vast distances between production and consuming centers, it was by far cheaper for Canada to sell oil to the United States and to import relatively cheap oil from other countries, particularly Venezuela. Furthermore, under this procedure Canada obtained additional benefits from the high prices prevailing in the protected United States market.

SUMMARY

In summary, it might be said that the 1957 Suez crisis was a warning to great industrial nations, as it demonstrated their vulnerability in depending to such a high degree on the oil supplied by a handful of countries. The closure of the Suez Canal interrupted the flow of Middle East oil to Europe, drastically changing transportation routes of crude toward the consuming areas.

It is essential to stress three fundamental facts related to this crisis: first, the vulnerability of the Suez Canal as a route of communication; second, the shortage of tankers, causing the steep rise in freights; and third, the increase in oil prices.

On realizing the significance of the first of these facts, consuming nations and oil companies started seeking sources of production that would lighten the risk of dependency on oil transported through the Suez Canal. Although this search was carried on throughout the world, it was felt most intensely in the African continent.

I have already mentioned the wide scope of activities started at the beginning of the 1950s by a group of independent United States oil companies. At the time of the Suez crisis these companies had an important volume of production which they were striving to place in the growing United States market.

I have also already referred to other factors causing havoc in the international oil market equilibrium. Greatly significant were the entrance of Russian oil into world markets, the reopening of the Suez Canal, and the introduction of the restrictive program covering oil imports in the United States.

The first two contributed to increasing oil supply, and the third restricted demand by the world's major consumer of hydrocarbons. Thus, excess supply was available, described in the oil argot as "oil surplus." This surplus negatively influenced prices, causing their decline.

In view of this critical situation, producing countries reacted by adopting measures that culminated in the creation in 1960 of the Organization of Petroleum Exporting Countries (OPEC).

SIX

The Era of Pérez Alfonzo (1958–1964)

INTRODUCTION—THE 1958 GOVERNMENT COUNCIL

In January 1958 Venezuela awoke to a new government and a new system. A luminous democracy had replaced its long night of dictatorship. During 1958 the destiny of the nation was ruled by a Government Council which, in the process of transition, had the fundamental mission of leading Venezuela toward a representative democracy and restoring the liberties of its citizens. The December 1958 elections were the culmination of this process, when the outstanding political leader, Rómulo Betancourt, was elected constitutional President of Venezuela, to take office in March 1959. During its thirteen months of life, the Council was headed by Rear Adm. Wolfgang Larrazábal and Dr. Edgard Sanabria.

In December 1958 the Government Council, which had inherited a series of political and economic problems generated by the dictatorship, conscious of immediate financial difficulties but probably unaware of external factors that were starting to negatively affect Venezuela's oil industry, decreed a tax increase on the oil companies' incomes, applicable from 1958. I shall refer to this tax reform in this chapter.

The oil companies operating in Venezuela were undoubtedly aware of the negative external factors that were beginning to affect the country's oil industry, and it would have been convenient to have them become further involved in the business, securing their continued investment and maintaining their maximum interest in Venezuela, in order to compensate—as much as possible—for such negative influences.

The fundamental problems the government faced were to prevent removal of oil companies to other competitive areas; to avoid paralyzation of investments in Venezuela, especially in the exploration field; to avoid the dismantling of oil fields; and wherever possible, to neutralize Venezuela's diminishing importance in international markets.

On the other hand, it was essential to protect the stability of oil prices. Any major production increase was certainly not recommendable at a time when production exceeded demand and the Middle East countries, with their extraordinary reserves, plus the newborn African oil industry, were swamping the markets. It was therefore essential to avoid production contests that would weaken and deteriorate international market prices, seriously affecting Venezuela's income. This was one of the main theses held for many years by the future Minister of Mines and Hydrocarbons, Juan Pablo Pérez Alfonzo, and his followers.

I must insist, however, that it was likewise necessary to avoid the industry's eventual deterioration in relation to investments, to securing reserves, to the development of new deposits, and to production capacity.

The Government Council's de facto administration (January 1958 to March 1959), lacked a petroleum policy. Apart from the December 1958 tax reform—the purpose of which was to accumulate fiscal resources, which were nearly depleted because of the enormous floating debt left by dictator Pérez Jiménez—the Council did not make any studies to determine the direction in which Venezuela's petroleum affairs should be oriented, then or in the future when a constitutional, democratic, and representative government was finally to be instituted. There were three reasons for this lack of concern.

The first reason is the brevity of the period in which the Council governed, i.e., slightly over thirteen months.

Second is the difficult political transition that the provisional government had to achieve, propitiating the fundamental change toward a system of democracy after the dark dictatorship that had prevailed throughout ten uninterrupted years. The system of political liberties introduced by the Government Council was continuously besieged by threats from sectors attempting to upset its stability. Throughout the Council's interregnum, it also experienced a series of executive changes at its top, medium, and even lower levels. Simultaneous to its political-administrative activities, it had to successfully prepare and carry out, in a very short time, general elections by means of popular, direct, universal, and secret vote, in which several political parties, still in various stages of reorganization and formation, participated.

The third reason for lack of concern is the fact that most of the political, military, and business leaders involved in this government lacked any knowledge of the oil business, a frequent phenomenon in Venezuela, and therefore they were unqualified to measure the depth of the clouds that overcast the international field and were soon to darken Venezuela's oil industry. Nor were these leaders capable of drafting any policy or strategy in relation to Venezuela's vast economic activity, its oil industry.

The 1958 Tax Reform

On December 19, 1958, the Government Council introduced a decree which, in addition to being aimed at increasing the state's participation in oil-industry profits, was intended to balance the 1958 and 1959 national budget. In 1958 the budgetary deficit was estimated at $580 million. Expenditures and investments, added to overdue and forthcoming payments for public works, amounted to $1.8 billion, whereas existing and foreseeable income and resources amounted to only $1.3 billion.

It was obvious that this deficit would carry into the following year, because even assuming that expenditures and investments would be maintained at $1.6 billion, as in 1958, liabilities amounting to $166 million would mature in fiscal 1959, in addition to a further $122 million corresponding partially to a foreign bank loan.

On the basis of these estimates, and by means of the aforementioned decree, the provisional government proceeded to reform the income tax law so that through new fiscal income it would be able to balance the national budget for those and future years. The reform did not affect the basic structure of the existing tax system, but rather modified the tariffs applicable to nonresident taxpayers and to fortuitous profits and, likewise, the progressive complementary tax tariff.

The application of this reform to oil companies and their activities was essentially because the change affected the progressive complementary tax tariff, and oil companies were the highest profit-making concerns in Venezuela. The first impact the decree had on oil companies is described in statements such as: "The drastic tax increase would adversely influence the competitive position of Venezuelan oil in world markets."[1] This opinion was stated in a letter dated December 22, 1958, from Harold W. Haight, president of Creole Petroleum Corporation, who expressed his surprise over the decree and requested a government reconsideration of same.

The Minister of Mines and Hydrocarbons, Dr. Julio Diez, in his reply dated December 23, 1958, expressed the following: "It is inadmissible to expect the government to reconsider the measures to which you refer;

[1]Letter from Harold W. Haight, president of Creole Petroleum Corporation, to the Minister of Mines and Hydrocarbons. Record of the Ministry of Mines and Hydrocarbons, Caracas, 1958, p. 369–370.

they were adopted by sovereign decision and subsequent to thorough and careful study. Such measures shall be upheld in their entirety."[2]

We can see here again that the traditional custom of improving the state's participation in the oil industry's profits by means of reforming the income tax law was being applied.

THE ERA OF PÉREZ ALFONZO

My sorrows are born from my philosophy . . . and I am more of a philosopher in prosperity than in misfortune.   SIMÓN BOLÍVAR
Letter to the Marquis of Toro, Chancay,
November 10, 1824.

In 1959 President Rómulo Betancourt appointed Dr. Juan Pablo Pérez Alfonzo as Minister of Mines and Hydrocarbons. Insofar as petroleum affairs are concerned, Dr. Pérez Alfonzo is foremost among Venezuelans, having been publicly and privately active in Venezuela for the longest period of time. For this reason I shall refer to him very especially herein.

After the death of Juan Vicente Gómez, Pérez Alfonzo began expounding his ideas concerning the nation's oil policy in the National Congress, representing the sector opposing the governments of Generals Eleázar López Contreras and Isaías Medina Angarita. When the latter was removed from office in 1945, Pérez Alfonzo became Minister of Development under the military-civilian junta. He then held the same position under Rómulo Gallegos, constitutional President elected in 1947.

From October 1945 to November 1948, Venezuela's petroleum policy was based on Pérez Alfonzo's opinions. However, after Gallegos was overthrown by the 1948 military coup, Pérez Alfonzo's theories lost influence in the orientation of Venezuela's hydrocarbon policy and were not to reappear until 1958, the year ending Marcos Pérez Jiménez's dictatorship.

With the advent of the constitutional democratic administration in 1959, Juan Pablo Pérez Alfonzo reappeared in government as Minister of Mines and Hydrocarbons, and again his ideas profoundly orientated Venezuelan petroleum policy. Since then, Pérez Alfonzo's concepts and outlooks have weighed significantly in national public opinion and policies adopted by Venezuelan governments, including President Rafael Caldera's Social Christian government. Therefore this stage in Venezuela's

---

[2]Letter from Dr. Julio Diez to the president of Creole Petroleum Corporation, in the Record mentioned above.

petroleum history can quite appropriately be called "the Era of Pérez Alfonzo."

Some of the most outstanding aspects of Pérez Alfonzo's activities during the period in which he was the Minister of Mines and Hydrocarbons (March 1959 to December 1963) involved: the policy of no further concessions; price protection; activities aimed at obtaining hemispheric preference for Venezuelan oil; creation of the Venezuelan Petroleum Corporation (CVP); the 1961 tax reform; and the creation of OPEC, Venezuela being one of its principal promotors. I feel that a critical examination of these outstanding activities and an analysis of his personality are essential in order to grasp the scope and recent consequences of Pérez Alfonzo's policy and influence in relation to the development of Venezuela's oil industry.

Parallel to some of his conclusions and deeds, such as hemispheric treatment and creation of the Venezuelan Petroleum Corporation, which undoubtedly benefited Venezuela, and others which benefited all oil-exporting countries, such as price protection and the creation of OPEC, his policies and teachings also contributed to a progressive weakening of Venezuela's oil industry as a direct result of the process of noninvestment and gradual decapitalization by concessionaires, to which the state did not respond by introducing any substitute compensatory measures.

This weakening did not become apparent for some time due to a series of circumstances: e.g., Pérez Alfonzo's accuracy in some of his projects, theories, and even predictions, such as the oil shortage; the nationalistic zealousness that he awakened, capably and tenaciously showing how companies operating in Venezuela had been overly favored in the past; the increase in fiscal profits per barrel since the mid-1960s, reaching exceptional heights from the beginning of the 1970s; the growth of world markets; the lack of oil knowledge by most Venezuelan leaders and by the populace in general; and the fact that Venezuelan governments have the tendency to give the nation the impression that all is well in its vast oil industry (considered fundamental for creating a sense of general confidence).

All these factors have contributed to Venezuela's not having, even today, a clear concept of the process that has gradually and quietly been developing since the beginning of the Pérez Alfonzo era: the decline of its oil industry. This process is so far advanced today that to reverse it looms as a mighty feat. Be that as it may, Venezuelans must face it in its true light, without extenuations, without unrealistic optimism, for the purpose of reaching a clear diagnosis of the country's oil situation today. This is the only way in which the country will be able to make wise decisions, based on the elaboration and execution of a full-scale program strategically effective and destined to reactivate its petroleum potential so that it might once again, and for many years to come, belong to the group of nations possessing a prosperous and expanding oil industry.

The "No More Concessions" Policy

Pérez Alfonzo, from his seat at the Ministry of Development from 1945 to 1948, had already defined the government's negative attitude in relation to granting new concessions. However, in those days, due to the temporary nature of the military-civilian Council's de facto administration and to the change toward representative democracy characterized by Rómulo Gallegos's government, as well as to the ephemeral nature of both governments plus numerous other problems, Pérez Alfonzo was unable to fully apply his plans concerning petroleum affairs.

But from 1959, and notwithstanding the dangers gathering over Rómulo Betancourt's new constitutional government, Pérez Alfonzo had a firmer political base and had acquired a more solid oil knowledge from his maturity in exile. He then launched the watchword of "No More Concessions" by means of several public proclamations through the press, radio, and television, and at meetings and conferences. The constant hammering of this negative attitude concerning concessions then became known as the "No More Concessions policy" of Pérez Alfonzo and the Venezuelan government.

The occasions are countless, especially during 1959, 1960, and 1961, when Pérez Alfonzo, acting intensively in his capacity as Minister of Mines and Hydrocarbons,[3] decried this position, which we might well classify as a watchword. He resorted to repetition—a fundamental instrument of proved effectiveness in the field of propaganda. It was through this medium that he planted in the minds of most Venezuelans, as usual oblivious to the activities and fluctuations of their country's petroleum affairs, the idea that concessions were an element negative and adverse to the nation.

Through logical inference, those who embraced this thesis as a result of the repetitious "No More Concessions" cry of the Minister, possibly reached the conclusion that all concessions (including those prevailing) were pernicious. In other words, if it is bad to grant new concessions, then, by association of ideas, existing concessions must also be bad.

I do not wish to contend that it was erroneous of President Betancourt's constitutional government, or of subsequent governments, to stop granting new concessions. Quite the contrary; I feel that new ways should be

---

[3]Although he was still Minister of Mines and Hydrocarbons in 1962 and 1963, he spent most of those years directing oil policy from his private home.

sought, now, to secure the state's effective and total control over its oil industry. I nevertheless think that this rallying cry by the topmost government officer—after the President of the Republic—was self-defeating, as he was thereby contributing to weakening the confidence of concessionary companies at a critical moment in petroleum exportation and when Venezuela was not, as it is today, in a position to nationalize its oil industry.

This I feel was detrimental to Venezuela, because its oil industry represents an extremely complex economic activity. A policy should have been adopted based on a dispassionate analysis of its various aspects and using all appropriate means to favor the interests of the nation and of its people over the short, medium, and long terms. It is my opinion that it was imprudent to have created a state of national emotion against those who were handling Venezuela's oil business. Venezuela's oil industry was in dire need of a wide and effective policy aimed at attaining its expansion and the state's capable, factual, and unemotional control over it. It needed a policy able to counterattack the industry's ailment since the beginning of Pérez Alfonzo's era: its decline.

On the other hand, upon ruling out one alternative, upon definitely blockading the road that was fundamental to the development of Venezuela's oil industry—the system of concessions—the immediate introduction of other means obviously became essential so that this basic activity could continue being channeled and developed efficiently to the benefit of the nation. However, Pérez Alfonzo was quite satisfied over having closed that door; he repeatedly announced that it was not to be reopened. He did not establish any system to substitute for concessions and permit the industry to continue expanding through new investments and adequate exploratory and development programs.

The Venezuelan people in general—overwhelmingly unaware, I repeat, of petroleum affairs—did not react in the face of this new situation. I am barely able to mention one isolated voice, that of Arturo Uslar Pietri, a prominent writer who is also fully conversant with economic affairs, who persistently and publicly clamored for a policy that could appropriately be substituted for concessions.

This reaction by Uslar Pietri, passively supported by many Venezuelan leaders, finally led, in 1961, to the announcement by Pérez Alfonzo of a substitute system, i.e., service contracts. However, and as it was later to be proved, the Minister was not overly interested in putting this system into practice, although he had outlined it and it was already being applied, on a small scale, in other exporting countries. Apparently by announcing the idea he was merely seeking a way of sidestepping the well-founded criticism voiced over his petroleum policy and over his not offering any alternative.

It was only after Pérez Alfonzo left the Ministry that, very slowly, a basic study of this new contracting system was initiated. The new system was finally put into practice many years later under President Caldera's administration. The bases of service contracts were approved in 1969, and the contracts were signed in 1970. They were then strongly attacked by Pérez Alfonzo.

Hemispheric Treatment

The so-called hemispheric treatment—or hemispheric preference—is a very important aspect of Venezuela's international oil policy, tending to secure adequate participation in the United States' higher demand for energy. However, this political thesis never received the expected acceptance in the official spheres of the United States, and Venezuela's reiterated insistence went unheeded.

In his book *Venezuela: Policy and Petroleum,* Rómulo Betancourt mentions that on April 24, 1959, his government delivered a memorandum, approved by the Council of Ministers, to the U.S. State Department. One month later, May 25, 1959, the Ministry of Foreign Affairs delivered another communication to the United States Embassy in Venezuela, as announced two days later by Pérez Alfonzo in the Chamber of Deputies. Both documents contained the government's findings concerning the oil-import restrictions ordered by the United States government.

The first communication outlined the government's position in relation to the discriminatory measures, whereby exceptions being granted to Canada and Mexico did not extend to Venezuela, stating that the restrictive measures "should have at least given preference to the Western Hemisphere."

President Betancourt's government (as well as successive governments right up to 1970, when this policy became inoperative because demand exceeded supply) tenaciously maintained its request for hemispheric treatment, consisting of extending the same preferential terms granted to Canada and Mexico to all Western Hemisphere oil producers. This position was stressed at meetings held between Presidents Betancourt and Kennedy (in Washington and Caracas), Presidents Leoni and Johnson (at Punta del Este, Uruguay), and Presidents Caldera and Nixon (in Washington).

In this respect a very important role was played by the Venezuelan Embassy in Washington. Many were the meetings held at technical and political levels by Venezuelan and United States officials. However, the discriminatory treatment prevailed throughout the entire decade of the 1960s in spite of the arguments and voluminous documentation submitted by Venezuelan governments. These arguments were based on the facts that Venezuela was the traditional and principal oil supplier for the

United States in times of peace and war; Venezuela maintained a higher degree of political stability than nations situated in the Eastern Hemisphere and was geographically closer to the United States; it had important oil reserves (at least up to some years ago); and it was one of the major importers of products manufactured in the United States.

The Situation Today

Although chronologically this point goes beyond the scope of this chapter, I feel it is recommendable—for the benefit of consistency and for better comprehension—to briefly refer to the situation as it stands today.

At the end of 1972 and beginning of 1973, the energy situation in the United States became particularly critical when its own internal oil production, plus imports, did not cover the high seasonal demand (winter). Under such circumstances, it appeared that hemispheric-preference measures (i.e., those granting exclusive benefits to Mexico and Canada) covering imports of heating fuel were no longer necessary, and the measures prevailed only until April 1973. In January 1973, United States import quotas for the same product were also temporarily suspended. Restrictions relating to maximum sulfur content in imported fuel were also eliminated. On the other hand, oil-rationing measures were introduced, and finally free imports were permitted.

Numerous United States sectors voiced their opinion concerning the advisability of allowing free play between supply and demand for setting the internal price for oil. Many of them, especially the local producers, had strongly opposed this policy for many years. Among the many arguments wielded against the hemispheric treatment, it is worthy to mention the one maintaining—quite logically as facts were later to prove—that traditional oil suppliers to the United States, namely Canada and Venezuela, could not keep up for much longer a rate of production growth sufficient to guarantee the supplies needed by the United States market.

Canada supplies nearly 50 percent of United States imports of crude, but in the face of the sustained growth of the United States market, experts believe that Canada has reached its maximum export limits. Canada's own needs are rapidly increasing, and scarcity of crudes from OPEC countries has made it difficult to maintain its former policy of importing cheap oil and exporting increasing quantities to the protected United States market at higher prices.

Venezuela's position has been widely discussed at all levels of public life, and I have frequently referred to it in this book. Its production capacity is tending to shrink as a consequence of hardly any exploratory activities for more than a decade. This phenomenon, added to the increased exploitation of its deposits, is causing a sharp drop in proven oil reserves. This, in turn, is translated into the relatively low capacity of Venezuela's oil production. It is thus that Venezuela today has lost command and hence no longer needs to demand, as in the past, hemispheric treatment in the United States import policy.

The facts that I have just outlined explain the reluctance of the United States (and also that of other industrialized nations) on extrahemispheric preference. The United States government is today preoccupied over its imminent dependency on Eastern Hemisphere sources, principally the Persian Gulf countries. These politically unstable countries have contracted a great volume of their production to other industrialized nations, such as Japan and those of the European Common Market, therefore forcing the United States to compete for its oil purchases.

It has furthermore become evident that the dependency of the United States (and also that of other industrialized nations) on extra hemispheric oil is being used as a political weapon by Arab countries, which are attempting to force United States decisions on several international issues in different fields.

Finally, due to the substantial growth of the internal needs of the United States, in April 1973 President Richard Nixon suspended direct control over imports and established a license-free quota system to protect home producers from foreign competition, thus virtually freeing the entrance for petroleum imports.

THE VENEZUELAN PETROLEUM CORPORATION

To be, comes first; the way of being comes next.

SIMÓN BOLÍVAR
Letter to General Francisco de Paula Santander,
Cuzco, June 28, 1825.

The Venezuelan Petroleum Corporation (Corporación Venezolana del Petróleo, CVP) is the state company that was constituted on April 19, 1960, for the purpose of directly exploiting hydrocarbons. Many countries, oil producers or not, have organized state companies for directly exploiting the different phases involved in the industrialization and commercialization of hydrocarbons.

Along political lines the creation of these companies, regardless of their form of organization, represents one of the means used by the state for securing direct intervention in national petroleum affairs. This managerial state function is taken from a philosophical concept totally opposite to the economic and political freedom based entirely on private enterprise;

however, such intervention is considered by many as necessary and vital to the state for security reasons and because of the economic importance of the oil industry proper and its incidence on national life.

State intervention is carried out either through the creation of official bodies for regulating production, through restrictive trade laws, or finally, more directly through the formation of state oil companies. In the United States, for example, the organizations established for regulating the industry are the Texas Railroad Commission and the Interstate Oil Compact Commission, and among the restrictive laws applicable to the oil business are the Sherman Antitrust Act, the Connally law, and the Clayton law.

The following are some of the state oil companies created in Latin America and other areas of the world: in Argentina, Yacimientos Petrolíferos Fiscales (YPF); in Bolivia, Yacimientos Petrolíferos Fiscales Bolivianos (YPFB); in Mexico, Petróleos Mexicanos (PEMEX); in Chile, Empresa Nacional de Petróleo (ENAP); in Colombia, Empresa Colombiana de Petróleo (ECOPETROL); in Brazil, Petróleos Brasileños, S.A. (PETROBRAS); in Italy, Ente Nazionale Idrocarburi (ENI); in France, Compagnie Française des Pétroles (CFP); in England, British Petroleum (BP)[4]; in Iran, National Iranian Oil Co. (NIOC); in Kuwait, National Petroleum Co. (NPC).

Economic and security reasons (besides the ancient greed for power by many leaders) and the worldwide importance of petroleum—the most commercialized element in the history of mankind—induced many countries to create state-owned companies. However, while this policy had already become factual in the oil world, in Venezuela it was still a mere manifestation of ideas, a phenomenon drawing the attention of foreign observers.

In a country such as Venezuela, the name of which is automatically linked with oil through association of ideas, neither nationals in the private sectors nor the state had any direct participation in the country's oil business. The absence of Venezuelan entrepreneurs is explained to a certain point by lack of great private capital in the country; however, the absence of the state, in a country which since 1928 had achieved notoriety as an oil producer and exporter, was incomprehensible to both Venezuelans and foreigners.

Argentina criticized Venezuela for not having created a state-owned company. In 1938, the Buenos Aires *Oil Information Bulletin* published an article entitled "Oil in Venezuela," which mentions the direct exploitation exercised by the Argentine state through Yacimientos Petrolíferos Fiscales, and proceeds to compare Argentina with Venezuela's case in the

[4]BP includes private capital.

following terms: "Venezuela could have adopted a similar procedure, even obtaining more favorable economic results, considering the higher production quality of its rich deposits which are by far more prolific than our own."[5]

Voices were also raised in Venezuela, clamoring for the creation of a government company for the purpose of directly exploiting the nation's hydrocarbons. In 1946 a committee was appointed for the "study and formulation of a project aimed at initiating for the first time a system different to that of concessions, for preserving the feasibility of diversifying the risk through participation by private investors."[6] However, nothing to that effect occurred.

Without entering into a deep analysis of the economic and legal factors pointing to the need for a state-owned organism in Venezuela, it is obvious that the creation of the CVP in 1960 represented a profound change in the nation's oil policy, which hitherto had depended on the traditional system of concessions and indirect exploitation. It is thus that the first step toward the stage of direct exploitation was taken to incorporate the state's own initiative in the development of the nation's oil industry.

With the exception of some opinions concerning activities carried out by this state-owned company in its thirteen years of existence, to which I shall refer later, its creation was a positive step in Venezuela's oil history, and there certainly is much that it can, and must, do in order to consolidate and expand its activities.

Aims of the CVP

The CVP was organized in the form of an autonomous institute,[7] attached to the Ministry of Mines and Hydrocarbons. Its main purpose, the reason for its existence, is to serve as an instrument for the government's oil policy. Article 3 of the 1943 hydrocarbons law establishes that the state is entitled to engage in activities related to the various phases of the hydrocarbon industry in two ways: directly through the nation's President, and/or through autonomous institutes or state companies. In other words, the specific functions to be performed by autonomous institutes or state companies are necessarily the same as those that the President should perform directly, according to the terms of the special law which provides

[5]Dr. José Ramón Ayala, *Epitome of Venezuelan Legislation and Mining Rights,* Tip. Americana, Caracas, 1945, vol. 11, p. 13.

[6]Dr. Luis González Berti, *Compendium of Venezuelan Mining Rights,* vol. 11, p. 386.

[7]Autonomous institutes are government organizations having their own legal authority and patrimony destined for specific activities assigned to them by law or statutes creating them.

for his intervention in this industry, of such vital importance to the nation's economy. In Venezuela, the state has predominant responsibilities in petroleum affairs not only involving the coordination and control over the industry proper, but also in its direct exploitation; not merely as a manifestation of policy by those exercising government functions, but also as a business activity.

According to the corresponding statutes in the hydrocarbons law, the specific field of the CVP embraces all phases of the oil industry. In effect, CVP is entitled to engage in all operations involving exploration, exploitation, manufacturing or refining, transportation, and marketing. Within its business powers CVP is entitled to promote oil companies with joint capital destined to industrial and commercial activities; to invest capital in same; and to subscribe or acquire shares, participations, or quotas in other companies having the same objectives. For the execution of these purposes, the statute empowers CVP to draw up agreements (i.e., service contracts) in general, for carrying out any activity that might assist it in reaching its goals.

CVP has not yet attained a degree of development consistent with its high objectives, because successive governments did not provide it with the necessary financial, technical, and political support. CVP possibly wasted time by allowing its expansion to depend on normal business growth, without developing an audacious program that would have led to its full participation in the oil business, in both the national and international fields.

This is also reflected in its balance sheets. During 1973, CVP's operations showed the following low results: Production of crude was approximately 82,500 barrels per day, slightly over 2 percent of the nation's total. CVP's refinery processed close to 20,200 barrels per day with a yield of 62 percent in Grade 6 fuel oil, 10 percent in Grade 4 fuel oil, 10 percent in diesel oil, 14 percent in gasoline, and 4 percent distributed among kerosene, gas oil, asphalt, LPG, and dry gas. Local sales of refined products in 1973 reached 41,600 barrels per day. Exports of refined products and crude oil reached 24,500 and 67,900 barrels per day respectively.

These figures indicate that CVP is not yet a first-rate oil company. In order to cover its domestic and international responsibilities, it is forced to resort to agreements with international oil companies operating in Venezuela, involving purchases and sales of crudes and by-products.

Time wasted in reaching the goal of transforming CVP into a first-rate oil company can be seen by the fact that it has not yet been able to supply the domestic market with gasoline and other by-products, legally provided for but in practice not attained. Although by law the right to supply gasoline and other by-products is reserved to the state, with control over the domestic market theoretically being in the hands of the state-owned company, in actual practice distribution, i.e., service stations, does not belong to the nation proper.

CVP's present condition, as a state oil company, is regrettable and directly in need of special attention and support by the government, not only to achieve major participation, but also because through nationalization of the industry CVP will be assuming direct control of a large proportion of the oil business. Hence its destiny will be to occupy a prominent position among the great international oil companies.

PRICE PROTECTION

Minister Pérez Alfonzo, as well as successive Venezuelan governments, paid special attention to the importance of safeguarding oil prices.

During President Betancourt's administration, a constant vigilance was maintained over prices for crude negotiated in the international market. Consequently, when the sharp decline in Middle East prices occurred in 1959, the initiation of which was attributed to British Petroleum, the Venezuelan government protested officially to the British government. Venezuela stated its position at international levels through official delegations, and together with four Middle East countries created the Organization of Petroleum Exporting Countries (OPEC), one of whose aims was to establish "the best means for safeguarding the joint and individual interests of member nations."

The coordinating Committee for the Conservation and Commerce of Hydrocarbons (CCCCH) was created in 1959 explicitly for protecting price levels established in trade agreements covering Venezuelan crude. In practice, this preventive function was able to paralyze all shipments involving prices not adjusting to the levels deemed acceptable by the government. Venezuela's activities in safeguarding prices were manifested through tax adjustments imposed on concessionaires, based on the government's official estimates of price levels at which exports of crude and product exports were to be made during the period 1957–1965, and through the creation of the aforementioned Committee. The government's estimates, higher than the levels used by oil companies in their tax declarations, gave rise to these tax adjustments.

Income Tax Law Reform of 1961: "Pay as You Go"

In general all Venezuelan income tax law reforms have been aimed at increasing tariffs applicable to the oil industry. However, the 1961 reform establishes that tax declarations, liquidations, and collections corresponding to mining, hydrocarbon, and other related profitable activities must be made in advance. In other words, the state receives such tax payments within the same taxable year instead of in the following year. This advance payment is then subject to final liquidation at the end of the corresponding fiscal year.

OPEC

Rather than being haphazardly arrived at, the Organization of Petroleum Exporting Countries (OPEC) was the result of an excellent idea of one particular person: Juan Pablo Pérez Alfonzo.

In addition to certain circumstances prevailing in the world's oil industry in marketing affairs, another factor strongly influential in OPEC's formation was the wish of all producing countries that exploitation of this perishable wealth be translated into the welfare of their people, and that it would become an efficient instrument through which they could rapidly overcome their status as underdeveloped countries.

Initially, this wish by oil-exporting countries to have an instrument of common defense was delayed in the Arab states, because their joint production in the 1940s and 1950s did not reach a percentage high enough to support their own decisions influencing the orientation of world oil policy, and they found it necessary to seek cooperation from nations other than the Arab world.

Some experts regard the so-called Oil Consultation Commission, which arose out of the First Congress of Arab Oil (April 1959) in Cairo, as being the prologue to OPEC. These experts mention Abdullah Tariki as being the leader of this project. Nevertheless, there is no doubt whatsoever about the prominent role played by Venezuela, through Juan Pablo Pérez Alfonzo, in originating OPEC, especially considering that previous intents by Arab states had only a regional or local scope, not worldwide such as OPEC.

The Arab League

We can see OPEC's first predecessor in 1944. On October 7, 1944, in Alexandria, representatives of a group of independent Arab states endorsed the so-called Alexandria Protocol, whereby the creation of the League of Arab Nations was proposed. In the following year, on May 22, 1945, seven countries endorsed the final agreement. In the text of this pact, Saudi Arabia, Egypt (later beoming the United Arab Republic), Iraq, Lebanon, Syria, Transjordan, and Yemen indicated the following main objectives as their purpose in signing the agreement:
- To strengthen relations between member nations
- To coordinate their respective policies so as to achieve mutual cooperation and safeguard their sovereignty and independency

■ To maintain a common interest in matters concerning the Arab countries

A Policy Committee was elected for coordinating these policies, and eight years later this Committee proposed the creation of a Committee of Oil Experts, which was in fact constituted on June 14, 1952, and annexed to the Policy Committee. We have here the first collective international body—which included countries that were not petroleum producers—established for the purpose of unifying some form of action to countereffect oil companies' monopoly over the exploitation of oil resources.

On January 20, 1954, the Oil Office was created by the Arab League, its purpose being to serve as consultant and assistant to the Committee of Oil Experts. In 1959 this Committee was transformed into the Department of Oil Affairs, although it remained under the Policy Committee. One decade later, in 1964, it was finally incorporated into the Economic Committee.

Possibly the most important event sponsored by the Department of Oil Affairs was the convocation of the First Arab Oil Congress, held in Cairo (April 16–21, 1959) and prior to the Baghdad Agreement which created OPEC. Venezuela was one of the countries subscribing to this agreement.

Creation and Activities of OPEC

The Organization of Petroleum Exporting Countries was established on September 14, 1960, in Baghdad, Iraq. Its name describes its essence: it is an international, intergovernmental organization, integrated by oil exporting countries. Originally (according to the Baghdad Conference), it was composed of only five members: Iran, Iraq, Kuwait, Saudi Arabia, and Venezuela. The following countries joined later: Qatar (1961); Libya and Indonesia (1962); Abu Dhabi (1967); Algeria and Nigeria (1972); and Ecuador (1973); in 1973 Gabon was accepted as an associate member (with no voting rights).

These OPEC member countries represent today the most significant bloc in the oil world, in which the world's greatest producing and exporting centers are united: Venezuela in the Caribbean Basin; the Middle East countries in the Persian Gulf Basin; the African countries; and Ecuador and Indonesia. Jointly they count upon the highest indexes of production, reserves, and exports. In fact, by 1973 OPEC countries were producing more than 55.4 percent[8] of the world's oil; 66.3 percent of international reserves was located in their territories; and they supplied more than 87 percent of the oil marketed in the world during that year.

[8]During the first half of 1974, OPEC's average oil production exceeded 31 million barrels per day, i.e., 55 percent of the world's total.

OPEC's Objectives

Clause IV of Resolution 1-2 concedes maximum importance to the aim of unifying the member countries' oil policies. It is thereby possible to unhesitatingly affirm that it is the objective of OPEC to exercise a certain influence in matters related to the oil industry in general, because however narrow a petroleum policy of any country might be, it nevertheless goes beyond price setting.

It is possible to extract other elements from the considerations and resolutions contained in the Baghdad Agreement, to describe OPEC's objectives. However, I feel it unnecessary, because an idea of OPEC's power as a representative instrument of oil-producing and oil-exporting countries lies in the mere fact that it is able to dictate and unify fundamental aspects of the petroleum policies of its associated member countries.

The preceding considerations concerning OPEC's scope and significance, based on the theoretical interpretation of its constitutional document (the Baghdad Agreement), already have a practical response in the field of relations between OPEC member countries and world oil companies.

OPEC was accepted as an international organization by the United Nations Social and Economic Council on June 30, 1965, and its participation as a representative organization of oil-producing and oil-exporting countries was also fully recognized and accepted by consuming nations and by the oil companies, especially after the decisions taken by the Organization at the XI Conference held in Caracas in December 1970. It might also be said that OPEC is already ahead of its initial objective, and its future depends on its capacity to continue consolidating and expanding.

OPEC's Head Offices and Organization

OPEC's permanent head offices were originally located in Geneva by decisions of the II Conference held in Caracas in 1961. However, at the IX Conference held in Tripoli in 1965, it was resolved to transfer them to Vienna.

The *conference* is the body exercising OPEC's topmost authority, and it is integrated by delegations representing the member countries. Each country designates a chief delegate, who is entitled to one vote at the conference. Decisions are approved by unanimity, except where concerning matters of procedure. The conference holds two ordinary meetings each year, although it is entitled to convoke extraordinary meetings when necessary. The meetings are held at the Organization's head offices, although they may be held in any of the member countries, or elsewhere.

The *board of governors* is integrated by governors designated by member countries and confirmed by the conference for two-year terms. It is

responsible for transacting Organization matters and for applying conference decisions. Its president, nominated by the conference, is elected for a
one-year term. The board of governors must also meet at least twice a year
at head offices or elsewhere.

The *secretariat* is the Organization's executive department and is subor-
dinate to the board of governors. It is integrated by a general secretary, an
under secretary, and other necessary personnel. It operates at OPEC's
head offices, assisted by administrative, economic, legal, information, and
technical departments.

The *general secretary* is appointed by the conference for a one-year term,
using the principle of alphabetic rotation. He resides at head offices and is
mandatorily a national of one of the member countries. He is the legally
authorized representative of the Organization.

The *Economic Commission* was created through decision reached at the
VII Conference, held in Djakarta (Indonesia) in November 1964. Its
objectives invest it as a special organization for the purpose of "assisting
OPEC in promoting the stability of international oil prices at equitative
levels." The Commission is formed by a board of national representatives
and by personnel belonging to OPEC's economic department.

OPEC Conferences

Throughout its entire existence up to January 1975, OPEC has held more
than forty meetings through its supreme body, the conference. At these
meetings numerous matters have been discussed and studied, and resolutions have covered many subjects, including: the statutes of the Organization (II Conference, Caracas 1961); modification to the statutes (VIII
Conference, Geneva 1965); transfer of head offices from Geneva, Switzerland, to Vienna, Austria (IX Conference, Tripoli 1965); declaration of
policies (XXXIV Conference); and others concerning petroleum policy
and production programs (XIV, XVI, XX, and XXXIV, all in Vienna).
Furthermore, since 1973, the Economic Commission has submitted documents for study by the conference related to oil prices, world inflation, net
profits by oil companies, and the situation of tanker freights.

OPEC in 1974 was composed of twelve member countries, as described
in Table 6.1.

OPEC's Oil Significance

World petroleum reserves on January 1, 1974, were 628 billion barrels, of
which 399 billion (nearly two-thirds) were controlled by OPEC's twelve
member countries. Insofar as world production of crudes is concerned,
during 1973 OPEC supplied an average of 30 million barrels per day, also
close to two-thirds of total production by non-Communist countries (45,-
900 barrels per day).

Table 6.1   OPEC MEMBERS

Country	Capital	Area in sq. miles	Population in thousands	Reserves Jan. 1, 1974, millions of barrels	1974 Daily production
Abu Dhabi*	Abu Dhabi	35,126	300	21,500	1,286
Algeria	Algiers	952,198	11,830	7,640	1,035
Ecuador	Quito	175,851	5,510	5,675	197
Indonesia	Djakarta	733,600	110,920	10,500	1,300
Iran	Tehran	626,200	25,780	60,000	6,000
Iraq	Baghdad	169,280	8,260	31,500	1,888
Kuwait	Al-Kuwait	9,375	468	72,750	3,144
Libya	Tripoli	679,358	1,740	25,500	2,117
Nigeria	Lagos	356,669	61,450	20,000	2,000
Qatar	Doha	4,000	75	6,500	555
Saudi Arabia	Riyadh	810,000	7,000	140,750	7,672
Venezuela	Caracas	352,143	10,720	14,000	3,370

*United Arab Emirates; also includes Dubai, Sharjah, Ajman, Ummadal-Qaiwain, Fujairah, and Ras al-Khaimah.

OPEC sold 55 percent of the United States' total hydrocarbon imports, i.e., 3.2 million barrels per day, of which Venezuela supplied 1.6 million barrels per day. Should the wish of the United States' President, Gerald Ford, be fulfilled, to the effect of reducing imports by 1 million barrels per day, and if this decrease is applied totally to OPEC countries, United States dependency on exports by OPEC would drop to only 10 percent.

In relation to the European Economic Community, in 1973 OPEC supplied more than 84 percent of their oil requirements; Japan, another exponent of industrailized oil-consuming nations, imported from OPEC more than 90 percent of its petroleum needs during 1973.

CRITICAL ANALYSIS OF PÉREZ ALFONZO'S OIL ACTIVITIES

Among all Venezuelans, Juan Pablo Pérez Alfonzo is the one who became most renowned in the oil world, because of his intensive activity over the years and his projection at an international level. I shall attempt to set forth a generalized critical analysis of his successes and failures, many of which are, of course, connected with Venezuela.

His public activities date back to the decade of the 1940s. For those who might wish to study his personality and his ideas concerning oil, I would suggest they take 1943 as a starting point, as this was the year in which his interventions in the Venezuelan Chamber of Deputies marked notable events, such as his firm and well-reasoned vote relating to the hydro-carbons law that was approved in that year. In fact the text of his vote clearly shows a series of aspects that were to characterize his future tendency: The careful selection of appropriate opportunities for entrenching an opinion or a position for the future; a thorough study of petroleum matters and his knowledge thereof; a frequently keen fore-sight of the reaction to be expected in each case from the oil companies, and his staunch attitude in confronting the companies' forebodings of alarming problems within the oil industry arising from producing coun-tries' new demands in relation to taxes and other fields; a sense of propaganda aimed at spreading his ideas; a nationalistically tinted stand touching the emotional chords and patriotic sentiments of many Venezue-lan leaders in the face of foreign economic domain of the oil industry; an analytical, philosophical, and moral attitude concerning the justice and/or injustice of Venezuela's petroleum position and the situation of the country in general, even at an international level.

All these circumstances surrounding Pérez Alfonzo's activities helped to convince his readers and listeners, in the great majority unfamiliar with the subject, of the truth contained in his expositions, as well as helping them to understand his severe criticism and unfriendly stance toward concessionary companies. He stated:

> If there is something absolutely undeniable, as we can plainly see, it is the situation created and maintained in Venezuela by the oil companies. The manner in which they have exploited that wealth belonging to the Venezue-lan people, even though aware of the people's needs, is a public and noto-rious fact. Exploiting the weaknesses of those who, with and without right, have represented the nation, acting beyond the margins of rights and justice, these companies obtained illegitimate profits and caused tremendous ills that cannot be erased by merely introducing a simple legal clause in any law; there does not, and cannot, exist any legal design that can right a wrong.

The 1943 Venezuelan hydrocarbons law established that all imperfec-tions which up to that date had affected the legality of some hydrocarbon concessions granted under preceding law were to be validated, as long as the concessionaires involved submitted to the new law. Pérez Alfonzo opposed inclusion of this legal consideration:

> The privilege intended by the project, upon establishing the relinquishment of all the nation's rights against the companies, would constitute a mere legal subterfuge, without any conscientious support by the people. Such an arrange-ment could never meet with voluntary general acceptance, because the decision is neither reasonable nor fair. The Venezuelan nation, which is aware of, and deeply feels, the despoilment of what in equity and justice is its right, cannot be satisfied with its relations' being simply adjusted to the oil concessionaires' convenience. Future justice and equity have their own foun-

dations and explanations and it is neither moral nor reasonable to expect that they serve at the same time as a means for canceling debts.

I have taken these same principles as a basis for securing the nation's rights to receive reparation for the damages it has suffered at the hands of concessionaires, independently from fundamental wrongs of varying degrees that might be shown in headings under which they are assumedly protected. Fraudulent deceit, bad faith in compliance, the lack of foresight, and other wrongs are general defects contained in those contracts, as well as the eivdence that they concern exploitation of public assets producing, against public interest, exhorbitant profits by far exceeding normal limitations.

Pérez Alfonzo was to dwell on this subject again and again, taking advantage of every single opportunity that arose. Twenty years later, during the 1960s, and then again in the 1970s, he used more or less the same expressions: the plundering of the nation's interests by concessionaires; the exploitation of the nation's wealth; excessive income earned by the companies; their illegitimate profits. He of course added other harassments, because time plus experience gained, and his capacity for study were to enrich his knowledge and thereby his arguments.

The triumph in practice of some of his viewpoints and projects, the errors incurred by concessionaires, the Venezuelans' ignorance of petroleum matters, and a series of other personal and social-political circumstances were to determine the influence of Pérez Alfonzo's theories over several leaders, especially of the left, and the orientation and handling of oil affairs by democratic governments, not only within his own Acción Democrática party, but also within the Social Christian party when it gained power.

His attitude and opinions were therefore long-standing and well known to the oil companies. The oil companies did not entertain the slightest doubt concerning Pérez Alfonzo's negative attitude insofar as they were concerned, or regarding his personal qualifications, which would lead him to eventually become one of the most outstanding proponents of the anti-capitalist petroleum theory.

In 1945, when the Acción Democrática party came into power, Pérez Alfonzo took over the management of oil affairs through his post as Minister of Development. However, Acción Democrática's stay in government was transitive and tumultuous, and hence so was Pérez Alfonzo's. Therefore, at that time he was unable to leave a deep impression in the development of Venezuela's oil industry. However, the December 1945 decree by the Government Junta and the 1948 income tax law reform, aimed at increasing the nation's income through 50 percent of net profits derived by the industry, were undoubtedly a consequence of his influence, and he thereafter implanted in subsequent Acción Democrática governments the position of no more concessions.

In the face of such measures, and for two principal reasons, the concessionaires' confidence nevertheless remained intact. The first reason was the extraordinary expansion of the world market, starting when the Second World War ended in 1945 and barriers limiting free circulation of oil were lifted. The expansion proper of the United States' motorized economy had become notable, especially during the last three years, when the United States had joined the war. The rapid growth of energy consumption in the United States increased that country's need to import from Venezuela, where, furthermore, United States companies had already monopolized most of the production, thus transforming the United States a few years later (1948) into a net oil importer.

The other even more important reason was the short term for which Acción Democrática then governed in Venezuela (October 1945 to November 1948), during which popular elections took place, changing Rómulo Betancourt's de facto administration into a constitutional one presided by Rómulo Gallegos, Venezuela's renowned writer. Thus, a constitutional and democratic government—the only system in which Pérez Alfonzo could have exercised his influence—was in those days unable to become asserted or survive. Nor did conditions then prevailing permit other countries to raise their flags of nationalism in order to confront the great oil consortia dominating production, refining, transportation, markets, and techniques.

In such a state of affairs, the oil companies felt secure against this new contender, whose ideas and animosity toward them were well known. During those three years international conditions were better than ever for the industry. In Venezuelan internal affairs, and for the first time in its oil history, a popular party and a Minister of Development (Pérez Alfonzo) hostile to the companies were directing petroleum policy; however, the government was not sufficiently stable, and this was understood by those who analyzed Venezuela's political situation. Pérez Alfonzo had not yet become an adversary sufficiently strong to fight the oil companies, and therefore their position in Venezuela, instead of weakening, became even stronger after November 1948, when Rómulo Gallegos was deposed. Consequently, they were confident in national and international perspectives.

Ten years in exile awaited Pérez Alfonzo. This negative circumstance, however, proved to be highly important later in his career. Through the discipline and austerity that characterized him, he was able to profoundly study Venezuelan and world oil affairs during his exile in Mexico. a propitious center for the subject, where voluminous publications and vast intellectual activity prevailed. Furthermore, by then Mexico was an oil nation that had already lived through a threefold experience: the regime of concessions granted to the same oil companies that were operating in Venezuela; their expropriation; and direct oil exploitation by the Mexican state. This enabled Pérez Alfonzo to complete, as it might be said, his higher studies in such a complex subject and to reflect profoundly on them. Ten years later he returned to Venezuela with ample knowledge and the necessary maturity.

Pérez Alfonzo had to wait until 1959, after dictator Pérez Jiménez had been removed and a definite form of democracy became instituted in the country. Acción Democrática once again returned to power. The new government assigned the portfolio of mines and hydrocarbons to Pérez Alfonzo, who this time managed Venezuela's petroleum policy with due support and necessary stability.

At this time internal negative factors, which had been accumulating in the stage of expansion immediately preceding, reached a crisis, weakening the oil companies and the industry proper and contributing to helping the new Minister of Mines and Hydrocarbons develop his theories, his programs, and his plans, in both the national and international fields. Thus, in 1959, the era of Pérez Alfonzo commenced and thereby his influence over the sectors forming public opinion.

It is interesting to study some facets of Pérez Alfonzo's personality, taking into account his influence over Venezuelan oil policy as well as over the policies of other important export countries in recent years, in addition to his influence over various public leaders. Furthermore, it is possible that at the present time the future of Venezuela's oil industry may be decided—and this, neither more nor less, is what Venezuela's present leaders are facing.

There exists a current among Venezuela's intelligentsia, especially the political, which apparently follows Pérez Alfonzo's postulations. For this reason I think those postulations should be closely analyzed and studied. They have resounded—and still do so—in Venezuelan petroleum thought.

There are many questions to be answered. How did he achieve this influence? Are his postulations applicable in our present times? What is their degree of precision? Are they appropriate or inappropriate for Venezuela? What might be pursued and what should be disregarded? Why? In the course of this book I shall endeavor to answer these questions, but let us now return to Pérez Alfonzo's personality.

Several personal qualities aided Pérez Alfonzo in achieving his purposes. In the first place is his intellectual capacity and formation. His legal background, the practice of his profession as a lawyer, and his experience as a professor at the Central University of Venezuela, heading the School of Civil Law, added to a penetrating analytical disposition and enabled him to develop his "oil dialectic" and his litigant attitude in confronting his main contenders, the international oil companies.

The second quality is his inborn propagandist facility, his sense of publicity. Since his first steps into the oil field, years before the 1940s, Pérez Alfonzo always ensured that his opinions, outlooks, principles, theses, and projects should resound in public opinion. During his activities while heading the Ministry of Mines and Hydrocarbons, he maximized the intensity of his statements through all media. He established the practice of weekly press conferences, and to that effect held meetings in his office with journalists over several years, making statements and replying to questions concerning oil affairs, up to a point where such meetings, and his statements, became a deeply ingrained habit with journalists and the readers of Venezuela's press.

He frequently gave speeches and held conferences at universities, associations, and other organizations; he appeared on television; he granted interviews to foreign newsmen and thus became, for a time, practically the sole petroleum voice of Venezuela. In this manner he created his reputation as a petroleum expert, a great nationalist, and a man highly successful in confronting the oil consortia inside and outside of Venezuela. After leaving the Ministry Office in 1963, he continued projecting himself to the public, alternating long periods of silence, as happened throughout the Presidency of Raúl Leoni (1964–1969), with more or less frequent press conferences held in his private residence.

In the third place, his early, self-imposed, avid dedication to studying and investigating petroleum matters placed him immediately at an exceptional level in comparison to the overwhelming majority of Venezuelans, who were not interested, neither then nor now, in this nationally vital subject. His staunch willpower, and especially his tenacity, enabled him to become a specialist in the field of oil economics.

This extraordinary interest in the nation's political life formed this Venezuelan politician in a new pugnacious cast. Pérez Alfonzo has not always been a cool and dispassionate thinker, only interested in seeking out the truth—in this case, on oil—wherever it is to be found. He has, rather, been a capable student and investigator, always interested in demonstrating his preconceived theories and in achieving through consummated dialectic, definite, inexorable, and indispensable aims: the destruction of the international consortia's power in Venezuela and the implantation of an austere and puritanical political-economic system in the country.

It is thus that those who have followed Pérez Alfonzo's vital trajectory might well say that he was not a thinker who, like the typical Frenchman in Salvador de Madariaga's sharp essay "Spaniards, Englishmen and Frenchmen," acted first so as to analyze later. Rather, he was like the Englishman who thought first, in order to adequately channel his acts. That is the method Pérez Alfonzo pursued.

To exhaustively analyze Pérez Alfonzo's thoughts on petroleum, through his published works,[9] would be an extensive task. Apart from his

---

[9]His books published to the present include the following: *The Monetary Question, The Trade Union Clause, Venezuela and Its Oil, Outline of a Policy, Oil Policy, Oil: Essence of the Earth, The Tax Amendment, Public Expenditure and Oil, The Oil Pentagon, Oil Reserves, Service Contracts,* and *Oil and Dependency.*

printed work, however, his line of thinking can be seen by the extraordinary effects that his predictions have achieved on an international level. His line of thinking includes the following: His persistence in proclaiming and warning that petroleum is not a renewable asset; his insistence on the need to demand and to obtain a higher and fairer price for this product, pointing out the exaggerated profits earned by international oil companies; the application of conservative measures for exploitation of hydrocarbons; the need to conscientiously ingrain the fact that oil is a perishable resource belonging to the producing nations; the waste of oil and of income derived therefrom by exporting countries. These are concepts common to all nations blessed by nature with this extraordinary source of energy.

Then we have his recommendable strategy for all petroleum-possessing countries to obtain maximum benefits from their oil: Higher prices, nationalization or direct exploitation of hydrocarbons by producing countries, and finally, the need for these oil-producing and oil-exporting countries to become united in a common organism with a view to securing their rights to establish oil prices and regulate production. These are the aims, the practical measures, recommended by Pérez Alfonzo. Over the years they have become more defined in the policies of all countries possessing petroleum.

Pérez Alfonzo had been proclaiming this consciousness and petroleum policy throughout his life, especially in the last two decades, both nationally and internationally, and long before the advent of OPEC, since he himself contributed to its creation as a reflection of his own convictions. He was to wait many years before his ideas became ingrained in the oil world and before he succeeded in overriding adverse opinions. After OPEC was created in 1960, many were the commentators inside and outside of oil-producing countries who echoed the opinion of oil companies and industrial nations to the effect that OPEC was an institution condemned to failure.

However, in the 1970s many changes have taken place in the relations between oil-exporting countries and the international oil companies and consumer nations—years after the first evidence of Pérez Alfonzo's line of thought and actions—which have gradually and irrefutably proved that this Venezuelan politician had a clear vision of the possibilities and aims of OPEC, fundamental not only to Venezuela and all other oil-exporting countries in particular, but also to all countries of the Third World in general that are producers of raw materials.

The existence of OPEC, besides being beneficial to the Persian Gulf oil countries and to countries of other geographic regions, has become a beacon for guiding efforts of recovery aimed at advancing Third World countries, exploited consciously or unconsciously by rich nations. Thus, through Pérez Alfonzo, Venezuela has exercised leadership among world oil countries, carrying them to a foremost position in international economy. This leadership, for identical reasons, is applicable to countries producing other raw materials, and it might well—and must—be exercised by Venezuela, through it oil experience, among countries belonging to the Third World.

After this recognition of the importance of Pérez Alfonzo's activities at an international level, I shall now refer to their effects in respect to Venezuela.

It is interesting to see how Dr. Pérez Alfonzo, who with dedication and tenacity achieved so many goals in petroleum affairs, nevertheless contributed to restricting exploitation of the hydrocarbon industry in his own country. During his term as Minister of Mines, he allowed a yearly decrease in reserves to slide lower than production. He was complacent, tolerant, and passive in confronting this gradual shrinkage of reserves, which commenced when he took command of the petroleum affairs of his country in 1959.

In attempting to analyze the reasons that he might have had for seeking a decline in the oil industry and for efficaciously aiding that decline, we should take into account his point of view. For example, there is his animosity—in some cases based on sound reasons—toward those who had retained, and still retain, control over the oil industry in his country. It is difficult, however, to conclude that his animosity in this respect constituted his basic reason for accentuating restrictions on the oil industry.

The fact that control over the oil industry will pass to the state, either progressively or immediately in 1975, does not imply that the industry should necessarily decline. There are those who attribute Pérez Alfonzo's stated wish for the restriction and decline of Venezuela's oil industry as having its origin in the socialist tendency of this distinguished Venezuelan politician and oil expert. The argument of those who support this theory is that Pérez Alfonzo, through restriction of the industry, is possibly seeking a general decline in the Venezuelan economy, which in turn would hit the foundations of capitalism and open the door to socialism.

But the reason carrying most weight could be his wish to curb wastage, through restriction of the oil industry, and hence detain the impoverishment of the country caused by the squandering made possible by the oil bonanza. It would appear that Pérez Alfonzo, prompted by his emotions and reasoning, and overlooking the existence of other means—in my opinion more appropriate—such as teaching conservation and even imposing obligatory norms, might be seeking the extreme projection of his austere image in Venezuela.

To conclude this analytical criticism of Pérez Alfonzo's deeds and lines of thoughts where oil is concerned (and which for obvious reasons are of major practical importance in the formulation of an ample and coherent petroleum policy), it might be said that it is regrettable, contradictory, and negative that such an extraordinary effort as the creation of OPEC, to which Pérez Alfonzo, and hence Venezuela, contributed so decisively, is not being effectively taken advantage of by Venezuela, precisely as a sequel to the policy and teachings of this outstanding Venezuelan oil expert.

OPEC, from an economic point of view, is the most effective and notable creation ever jointly and successfully established in the history of mankind by subdued and exploited countries for protecting their interests. As a consequence of OPEC, and based on the worldwide critical oil shortage and the industrial nations' increased dependency on this fundamental element, a handful of oil-exporting countries, for the first time ever, have been able to impart their terms to the consumer nations. The respective incomes of this small group of countries are increasing enormously, and it is expected that this situation will continue for several years, thus enabling them to rapidly accelerate their development and expand their economic and even political powers throughout the world.

Venezuela, for many years the third oil producer and the first oil exporter in the world, up to a short time ago showed other oil-exporting countries how to handle their relations with exploiting companies. Mostly as a result of the policy applied by Pérez Alfonzo during the 1960s, when he was Minister of Mines and Hydrocarbons, and also under the influence of his teachings, Venezuela's oil industry is undergoing a process of decline. As a consequence, although Venezuela was the fundamental promotor of OPEC, its power within OPEC is also declining. Should this tendency persist, Venezuela will be unable to enjoy as it should the fruits of OPEC, together with the Eastern Hemisphere nations—the formidable benefits, both economic and political, arising from their respective expanding oil industries.

It is in this factual consequence that we can find the fundamental negative aspect of this outstanding Venezuelan public figure. Nonetheless, his action and philosophy on petroleum matters, in the midst of many achievements, helped save Venezuela's oil industry—and also that of other exporting countries—from the domination of multinational oil companies.

SEVEN

The Years of Oil Peace (1964–1970)

At the beginning of Chapter 5 we saw how several detrimental factors, accumulating at an international level during the 1950s, emerged toward the end of that decade to originate the so-called oil suplus, which in turn caused a slump in world oil prices. This adverse situation affected all export countries, particularly Venezuela—at the forefront in those days. From 1959, income per barrel started decreasing.

It is somewhat difficult to pinpoint the degree to which the income of Middle East countries was affected during this price recession. Before the existence of OPEC, Persian Gulf countries collected taxes from concessionaires on profits based on closing prices. In some cases, since 1956, companies granted discounts on those prices. After OPEC was constituted, these countries began calculating their taxes on the basis of posted prices instead of closing prices, and it is therefore assumed that they were thus partially able to counterbalance the price drop for fiscal purposes.

In Venezuela, tax computations were still theoretically based on closing prices (selling price less discount). However, in actual fact this was not the case, and from 1960 the Coordinating Committee for the Conservation and Commerce of Hydrocarbons (CCCCH), created the preceding year, started raising objections to prices applied on numerous transactions as excessively discounted and causing higher price levels for fiscal purposes. A tax redress finally reached in 1966 contributed to counterbalancing the lower income per barrel received by Venezuela during the period 1959–1965, as a consequence of the general price drainage. In any case, and although it is difficult to determine which export countries were most affected by this phenomenon, it is possible to state that all of them, including Venezuela, felt a decrease in income per barrel.

However, the situation in Venezuela concerning income from its petroleum industry was different from that of other export countries, in particular those of the Middle East, whose income constantly increased due to uninterrupted and rapid growth in production. The Mideast oil industry has in fact continued to expand, a phenomenon clearly shown by production figures of that area. Between 1959 and 1972 the increase was fourfold, from 4,594,000 barrels per day in 1958 to 17,122,000 barrels per day in 1972. Reserves in the same period, meanwhile, doubled, from 158 billion barrels in the year 1959 to 356 billion in 1972.

It might therefore be said that the oil industry in the Middle East (as well as in Africa and the Far East, but in less proportion), although affected by the price recession begun in 1959, continued developing rapidly during the 1960s and thereafter. This permitted these countries to benefit from the unusual market boom of ever-increasing prices and profits, the effects of which have been felt so notably in the 1970s.

On the other hand, in Venezuela this factor of international price drainage was accentuated by a series of internal factors, creating an industrial recession in production capacity and reserves, which still prevails. Since the mid-1960s, the decline in Venezuela's reserves has become more accentuated, even though the state's income per barrel has gradually increased.[1]

In December 1963 Pérez Alfonzo ceased being Minister of Mines and Hydrocarbons, with the drawing to a close of President Rómulo Betancourt's term of office. Betancourt, an outstanding Venezuelan politician and statesman, had by then achieved a fundamental goal in assuring the destiny of Venezuela's democracy: for the first time in Venezuela's history a freely elected President had completed his constitutional term. During his government, subversive right- and left-wing uprisings that swept the country and threatened to overthrow its newborn democracy were overcome or neutralized. The critical financial situation left by dictator Pérez Jiménez had been restored to a healthy state by 1964. Thus, Venezuela had finally reached a minimum degree of organization, in both its political and economic spheres. The bases had been laid for a future climate of internal peace and for further economic and social development. These circumstances, Pérez Alfonzo's exit from the governmental scene, and the cautious personality of the incoming President, Dr. Raúl Leoni (March 1964) contributed to setting petroleum policy and relations with concessionaire companies within a framework of moderation and tranquility.

Dr. Rafael Caldera succeeded Dr. Leoni in March 1969, after a brilliant

---

[1]In 1958 Venezuela received $1.00 per barrel. This income decreased to $0.87, $0.82, $0.86, $0.88, and $0.89 in the years 1959, 1960, 1961, 1962, and 1963. In 1964 it increased to $0.92 and maintained approximately that level until 1966, when the reference price system was introduced. Thereafter, state participation has continuously increased.

electoral campaign, defeating Acción Democrática's candidate, Dr. Gonzalo Barrios. Dr. Caldera's tenacity and political ability finally brought him success as a leader of the COPEI.[2]

President Caldera's first two years in office (until December 1970, when the period considered in this chapter ends) are also characterized by tranquility in petroleum spheres. Thereafter and, as we shall see, later, a sense of nationalism erupted in Venezuela and throughout the world, and for the first time in Venezuela's history the conduct of its oil policy was transferred from the executive branch to the legislative branch.

During the years following Pérez Alfonzo's exit from the Ministry, i.e., President Leoni's full constitutional term and President Caldera's first two years in government, petroleum policy and oil events lost the accelerated rhythm, publicity, and pugnacity that had prevailed under Pérez Alfonzo.

This was a calm and less conflictive period in Venezuela's petroleum affairs, in which relations improved between the government and concessionaires, with mutual understandings replacing hostilities. In this manner important agreements were made, including those concerning fiscal adjustments, reference prices, and desulfurization plants during President Leoni's government, and the service contracts signed in 1969 by President Caldera's government.

Although the more moderate and calm climate characterizing Venezuelan petroleum policy in those years propitiated important agreements with concessionary companies, which were convenient to the nation's interests, no important change occurred that would have permitted Venezuela's industry to shed its stagnation and reinstate a vigorous development. The industry continued its decline in proven reserves, petroleum deposits, and productive installations in general.

AGREEMENTS BETWEEN THE STATE AND THE OIL COMPANIES

The Income Tax Administration had been considering that the export sales prices of crude oil and by-products declared by the oil companies in their 1957 to 1965 financial statements were way under the price levels

---

[2]COPEI was founded in 1946 and is the second greatest political party in Venezuela. It is also called Social Christian party, Christian Democratic party, and Christian Democracy. It has acted within the opposition and in government. Originally, it was a rightist party due to the orientation of some of its founders, who were of a deeply religious formation. Where ideology is concerned, it follows the international Social Christian current, and in recent years it has progressively shifted to the left, today being a center-left party. According to its manifestation of principles contained in its program, COPEI is a nationalistic, multiclass, democratic, Christian, and anti-communist party.

attained on the international market. It is obvious that if oil companies declare their sales prices at less than those actually obtained on the world market, their profits decrease, to the detriment of the National Treasury upon assessing income tax to be paid by each company.

Based on these facts, in 1963 the Income Tax Administration drew up a claim against companies that had exported crude oil and derivatives in the preceding years and officially reestimated the value of these exports, using as sales prices those that it considered appropriate for the operations involved. This redress represented an additional global income for the National Treasury presumably exceeding Bs2 billion.[3]

On the other hand, the affected companies maintained that export sales prices for crudes and by-products were the best they were able to obtain on the market and that prevailing market conditions should be used as the basis for tax liquidations. In other words, they were rejecting not only the measure proper but also the basis used for price estimating. Thus, these fiscal measures and corresponding tax bills were appealed by the companies in legal procedures submitted to the Supreme Court of Justice and the First and Second Income Tax Courts.

The procedures were finally concluded through agreements reached between the state and the oil-exporting companies in February and June 1966, which jointly represented to the nation an additional income on the order of $259 million.[4] The writs of these agreements had the same motivations and a basically similar text, including officially estimated export prices, with correspondingly applicable fines, as well as claims by the state for other concepts.

I shall now refer to three of these agreements, especially in relation to their economic provisions.

1. Agreement with Creole Petroleum Corporation and Subsidiary or Affiliated Companies

The agreement with Creole Petroleum Corporation (February 3, 1966) also included its five subsidiary or affiliated companies: Compañía de Petróleo Lago, Creole Investment Corporation, Lago Investment Corporation, Standard Tankers Company Limited, and Esso Transport Company Inc.

The amount agreed upon was $121 million[5] payable in cash in three

---

[3]Juan Pablo Pérez Alfonzo, *Oil and Dependency,* Síntesis Dos Mil, C.A., Caracas, 1971.

[4] The Ministry of Mines and Hydrocarbons Memoir dated 1966 states: "In reference to the claims, a payment of approximately $259 million has been agreed." However, in his speech via the national radio and television network (October 6, 1966), Dr. Pérez Guerrero, then head of the Ministry of Mines, estimated the amount at $226 million.

[5]The fiscal investigation covered the periods 1951–1953 and 1955–1965 inclusive, representing a total amount of more than $287 million.

equal quotas, the first to be paid upon signing the agreement and the other two within nine and eighteen months, respectively.

2. Agreement with Compañía Shell de Venezuela and Subsidiary or Affiliated Companies

The agreement with Compañía Shell de Venezuela (February 3, 1966) also included another nine companies belonging to the Royal Dutch-Shell group: Venezuelan Eagle Oil Co. Limited, Shell Caribbean Petroleum Company, N.V. Curaçaosche Scheepvaart Maatschappij, The Shell Petroleum Company Limited, Shell International Marine Limited, Societé Maritime Shell, Shell Tankers Limited, Asiatic Petroleum Corporation, and The Colón Development Company Limited.

The amount agreed upon was $73 million[6] payable as follows: (a) $57 million in three equal cash quotas, the first to be paid upon signing the agreement and the other two within nine and eighteen months, respectively; and (b) $22 million to be used for public works to be executed by the Venezuelan government. This last amount was to be paid gradually, i.e., by canceling invoices submitted by the government covering evaluations of completed jobs.

3. Agreement with Mobil Oil Company of Venezuela and the Mobil group of companies.

The agreement with Mobil Oil Company of Venezuela (June 6, 1966) included the other companies of the Mobil group: Mobil Oil Corporation, Mobil Marine Transport Inc., Mobil Development Inc., and Compañía Anónima de Petróleos Socony.

The amount agreed upon was $12 million[7] to be paid as follows: (a) $11 million in three equal cash quotas, the first upon signing the agreement, the second on December 22, 1967, and the third eighteen months from date of the agreement; and (b) $1 million to be applied to payment of evaluations of public works to be executed by the Venezuelan government.

General Comments on these Agreements

According to the texts proper, the objectives pursued through these agreements were (1) total cancellation of fiscal liabilities resulting from the tax claims against the companies; (2) termination of lawsuits that had reached the courts, and (3) abandonment of suits in the stages of reconsideration or administrative processing.

The aforementioned amounts to be paid by Creole, Shell, and Mobil,

---

[6]The fiscal investigation covered the period 1954–1965 inclusive, and involved a figure exceeding $318 million.

[7]I do not have available the amount involved in the claims.

and their subsidiary or affiliated companies, totaled $212 million. Therefore the difference between this amount and the $259 million—which the Ministry of Mines and Hydrocarbons had estimated as additional tax income covering claimed export sales prices and other concepts—should correspond to agreements drawn up by the state with other oil-exporting companies for the same purposes.

These agreements, besides solving the matter posed by tax claims made against the oil companies on their 1957–1965 financial statements, made provisions for the then-current 1966 fiscal period. These additional provisions were:

1. That in no event would official estimates or claims be made on prices corresponding to the 1966 fiscal period.

2. That such prices would be established as follows: Should the average 1966 price be less than prices attained in 1965, the Tax Administration would assess a complementary tax for the applicable difference; however, should the average 1966 price be equal to, or higher than, that of 1965, tax would be calculated on the basis of the 1966 average, and no additional tax whatsoever would be applicable to those prices.[8]

In the text of the agreements, no reference is made to the quantities claimed by the National Treasury from each of the companies, and therefore there are still doubts as to whether or not the interests of the nation were fully and duly protected by these agreements. In order to judge whether the amount finally received by the Treasury was a reasonable and concordant settlement, it would be necessary to have the precise figure claimed. But, as mentioned, unfortunately the documents covering the agreements do not give this information.

REFERENCE PRICES OR REFERENCE VALUES

Before defining reference prices, I shall briefly describe some of the terms related to petroleum price setting, which is obviously fundamental both to companies that industrialize and trade in oil and to governments that need a term of reference to calculate taxes to be obtained from the exploitation of this perishable natural resource.

As we have seen throughout this book, the terms "posted prices," "liquidating prices," and "closing prices" are frequently used in the petroleum business. They have the following meanings:

---

[8]These provisions apply to Chap. III of the agreement with Compañía Shell de Venezuela, dated Feb. 3, 1967. The agreements with Creole and Mobil have, in a general sense, a similar objective.

The posted price is the value at which companies offer their crudes and by-products for sale at ports of origin. As this price is established by the oil companies themselves, it is subject to change and therefore is not a safe basis for estimating the true value of transactions or a uniformity among them.

The liquidating price was established for the purpose of determining the state's participation in concept of royalties. This price, based on comparisons between the prices of Venezuelan and United States crudes, has the disadvantage of being connected to prices prevailing in an area outside of Venezuela and over which, of course, Venezuela has no modifying influence. However, as the United States constitutes a protected market, the price level serving as a comparative basis for establishing the liquidating price of Venezuelan crudes had been much higher than actual sales prices.

The closing price is the truly representative trade value assigned to petroleum, as it is the price at which transactions are actually closed in the market. However, it is also susceptible to modifications by the oil companies through discounts granted on their sales, thereby reducing the true figure on which the state should assess income tax.

According to the foregoing, the oil companies were obviously able to influence the setting and alteration of the three types of prices, to the detriment of oil-exporting countries such as Venezuela. This situation led export countries to seek measures aimed at counterbalancing these negative effects. Venezuela reacted at first by objecting—for fiscal reasons—to the sales discounts granted by the companies, and then by establishing reference prices.

Generally speaking, reference prices are the values fixed by the Venezuelan state for exports of oil and derivatives for income tax assessment purposes. They are something like an official regulation of values for exports of petroleum or its derivatives. This regulating function by the state has been gradually accomplished through the so-called reference prices, established either through agreements or facultatively and unilaterally by the executive branch of the government.

The modality of these agreements between the state and the oil companies in relation to reference prices has been enforced in Venezuelan legislation under varying formulas. In the 1966 income tax law, article 41, single paragraph, the Executive is authorized to "draw up prior agreements with taxpayers [read "oil companies"] for up to five-year periods, in order to establish the basis on which values of export items or merchandise, at Venezuelan ports of shipment, shall be made."

Based on this legal provision, the Venezuelan state drew up agreements with the oil companies covering reference prices, the first being those contained in the documents concerning fiscal claims, signed in 1967 and applicable to the 1966 fiscal period. Subsequently, in the 1970 income tax law reform, the system of agreements covering reference prices was substituted by the present system under which reference values (as they are now called) applicable to exports of petroleum, by-products, and derivatives are unilaterally established by the Venezuelan state. I shall refer later to this price setting by the state.

DESULFURIZATION IN VENEZUELA

It is a known fact that air is a mixture of colorless gases containing approximately 78 percent nitrogen and 21 percent oxygen; the remaining 1 percent is formed by several gases (argon, carbon dioxide, helium, ozone, hydrogen, etc.) and water vapors. When other substances are added, the air becomes polluted and, in some cases, possibly noxious for living beings, especially mankind. The degree of danger depends on the types of pollutants and on their higher or lesser concentration, and also on the degree to which the human organism is able to tolerate these injurious impurities.

One of the main problems confronting large cities in the world today is atmospheric pollution, especially by sulfuric oxide that becomes separated in combustion of heavy industrial and domestic fuel oils. Investigations concerning environmental pollution and air purification have, in recent years, increased in many countries. As a consequence, several governments have adopted legal regulations for the purpose of reducing the sulfuric oxide content in crude oil and its derivatives.

In 1963 the United States Congress approved the Clean Air Act, and more than $20 million was allocated in the period 1968–1969 to such activities. In May 1966 New York City introduced a regulation establishing the maximum percentage of sulfuric oxide permitted in heavy fuels used in the metropolitan area. The regulation set forth percentages, applied in several stages, as follows: 2.2 percent from January 1967, 2 percent from April 1969, and 1 percent from April 1971. Similar measures were adopted by the states of New York, New Jersey, and Pennsylvania, as well as the District of Columbia. Furthermore, the April 1971 date set for the 1 percent application was advanced to October 1969.

Such measures directly affected Venezuela, because they were applicable to the Eastern oil districts of the United States, precisely the main market for Venezuelan oil, which has a high sulfur content (2.5 percent).

Unless Venezuela desulfurized its oil according to the new environmental antipollution regulations, the traditional United States market would be closed to it, or in any case, it would run the risk of a decrease in demand and a resulting price decrease, since most crudes imported by the United States from other sources had a lower sulfur content. Crudes proceeding from Libya, Algeria, Nigeria, and Indonesia are among those with the lowest. Desulfurization required special refining plants, capable of reducing the sulfur content that could not be separated through normal refining at the plants then existing in Venezuela. Therefore, its oil industry needed a solution to the marketing problems arising from the aforementioned United States measures.

In order to meet this situation, and through the initiative of a government preoccupied over this problem, on July 20, 1967, the Venezuelan Congress approved the special agreements law covering hydrocarbon desulfurization. The purpose of this law was to stimulate the establishment in Venezuela of hydrocarbon desulfurizing plants capable of reducing the sulfur content of its heavy fuel so as to meet international market requirements. The law authorizes the government to take any necessary steps in this respect and establishes the country's desulfurization policy.

Desulfurizing Plants (Amuay and Punta Cardón)

Protected by the aforementioned law, the Venezuelan government entered into two special agreements. One was with Creole Petroleum Corporation (January 1968) for the installation of a desulfurizing plant at the Amuay refinery (Falcón state), and the other was with Compañía Shell de Venezuela (May 1968) for the installation of a plant at the Cardón refinery, also in Falcón.

The incentives granted to these companies for installing the desulfurizing plants included amortization of investment over a five-year period and exoneration from taxes and import rights covering materials, equipment, machinery, installations, etc., destined for construction, repair, expansion, and operation of the plants.

The initial combined capacity of the desulfurizing plants was 150,000 barrels per day; expansion up to 1973 brought processing up to 400,000 barrels per day of fuel with an average 1 percent sulfur content, and sulfur recovered amounted to 720 metric tons per day.

Questioning the Creole Contract

The special contract drawn up between the Venezuelan executive branch and Creole Petroleum Corporation is at present being questioned. On August 4, 1969, Dr. Francisco Alvarez Chacín filed a petition of nullity of the administrative act, in the Supreme Court of Justice, based on the memorandum signed January 5, 1969, by the Minister of Mines and

Hydrocarbons (Dr. José Antonio Mayobre) and the Minister of Finance (Dr. Benito Raúl Losada) and by the president of Creole Petroleum Corporation (Leo E. Lowry).

Dr. Alvarez Chacín, on the one hand, and Doctors Mayobre and Losada, on the other, have publicly expressed in recently published papers[9] their viewpoints concerning the nullity or validity of that principle document of agreement.

As mentioned previously, notwithstanding the conflict concerning the legality of these desulfurization agreements and their contents, they did represent an act of incentive, a policy, a reactivation by the Executive in the phase of decline in which the Venezuelan oil industry happened to be, due to the factors I have mentioned previously.

The Supreme Court of Justice, the highest tribunal in the land and the entity engaged in the pertinent investigations, will now have the last word concerning this case.

SERVICE CONTRACTS

As a form of introduction I might mention that service contracts are one of the means validly used by existing and/or future autonomous institutes or state companies to fulfill their responsibilities in directly exploiting hydrocarbons.

The 1943 hydrocarbon law does not precisely define service contracts; it merely uses the words "agreements" and "contracts." However, the declaration of objectives in the 1967 reform, referring to those "agreements" or "contracts," does in fact generically call them "service contracts."

This term does not apply to the traditional concept of civil law, because even though such instruments might embrace forms or objectives regulated by common law, their legal nature applies to the field of public law. It is *sui generis,* special and exclusive to hydrocarbons. It is a contractual form created by the legislature especially for hydrocarbons, and therefore its comprehension and interpretation must leave aside, as far as possible, notions applicable to any other legal norms or principles. I reiterate that it is a special contractual form referring exclusively to hydrocarbons, to be used by the Venezuelan Petroleum Corporation (CVP) or any autonomous institute or state company, existing or to be created, destined for the direct exploitation of hydrocarbons.

The term "service contracts" embraces under that single heading jobs

---

[9]Francisco Alvarez Chacín, *Oil Secrets against Venezuela: A Case in Court,* Caracas, 1970. José Antonio Mayobre and Benito Raúl Losada, *Desulfurization in Venezuela: A Nationalistic Decision* Caracas, 1970. Both published by their respective authors.

or services rendered by contractors to autonomous institutes or to state oil companies. Article 3 of the 1967 reform establishes the content of its clauses. Parties to these contracts are individuals or corporations who assume thereunder the obligation to perform, for the aforementioned entities, jobs or services related to the exploration or exploitation of hydrocarbons.

General Aspects

The duration of service contracts is for twenty years or, exceptionally, thirty years. The contracts may cover—jointly or separately—exploration and exploitation activities. We can see that whereas in the system of concessions the duration was forty or fifty years, in service contracts it is twenty or, exceptionally, thirty years. The exploration period under concessions was three years, but under service contracts it may even reach five years.

Territorial acreage granted through concessions was limited by law (general plots, 24,710 acres; exploitation plots, 1,235 acres); under service contracts the expanse is established in the context proper of the contract involved.

A concession grants the beneficiary a real estate right for exploring and exploiting, and the beneficiary may sell, grant, transfer, or mortgage the concession. Service contracts do not grant any real estate rights over the deposits; they merely grant a personal right agreed upon between the contractor and an entity of the state, and that right cannot be transferred, encumbered, or attached.

A service contract covering exploration and subsequent exploitation specifies that a contractor may select 20 percent of the original acreage for his own exploitation, with the remaining 80 percent reverting again to the autonomous institute, or state company, party to the contract. In this respect, service contracts also differ from the system of concessions, wherein the concessionaire may select 50 percent for exploitation out of the area granted. Hence, service contracts must represent greater advantages to the nation than concessions.

Among the advantages gained through the system of service contracts, one is said to be the reduced operational risk of the public entity in the event it should have to perform the services contracted for directly, since exploration expenses are borne by the contractor until the stage of commercial production is reached.

According to the preceding considerations, the system of service contracts was obviously drawn up in order to improve the traditional system of concessions. In fact, the 1967 hydrocarbons law proper establishes that "the terms and conditions stipulated in each contract must be more favorable to the nation than those provided under concessions."

Since the 1967 hydrocarbon law reform, five service contracts have been granted in Venezuela.

On March 24, 1968, the Venezuelan Petroleum Corporation published the minimum conditions for service contracts, inviting interested parties to submit their bids covering services to be rendered under assignments 9 and 16, situated in the southern area of Lake Maracaibo, covering 617,-761 acres distributed in five blocks of 24,710 acres each. Either independently, or in groups, eighteen companies submitted eleven bids. Finally, service contracts were drawn up as follows: three with Occidental de Venezuela (July 29, 1971); one with Compañía Shell de Venezuela (July 29, 1971); and one with Mobil Oil Company of Venezuela (September 16, 1971).

An Opinion Concerning Service Contracts

I was a member of the Chamber of Deputies when the subject was debated, and I voiced my support of the system of service contracts because I considered that it filled a void in the petroleum policy that had prevailed throughout the preceding twelve years and that it was also an instrument for revitalizing Venezuela's oil industry. Hereunder are some of the opinions I expressed at that time.

> After the void that has prevailed in petroleum policy during the last twelve years, fulfilled only in a minimal proportion through the creation and activities of the Venezuelan Petroleum Corporation, we are at present on the threshold of what appears to be a new chapter, as pointed to by Senator Uslar Pietri, in pro of the development of Venezuela's oil industry. If such a phase is opening, then we shall truly be taking a new step, and therefore I feel we should concentrate with objectivity, courage, and sincerity, with a historic consciousness and love of country, to analyze what the objectives, aims, and paths to be followed—over the medium and long term—are and should be.
>
> A few moments ago in this Chamber, we were reminded that Venezuela, throughout its oil history and notwithstanding its reverses, has in the long run always managed to advance in this field. However, it was also recalled that our history has frequently been marked by improvisation, a lack of planification, and a lack of a clear consciousness of what national goals are and should be in this field.
>
> It is therefore recommendable for us to constantly bear in mind the worldwide tendency toward long-term planification. You are well aware, Gentlemen of Congress, that industrialized nations, and even some in the process of development, formulate their policies, plans, and economic and social programs, through so-called "prospective analysis," attempting to

shape, over thirty-, forty-, or fifty-year spans (a long period in the life of man, but brief in the life of a nation) what the fundamental outlines of their respective societies shall be, based on social-economic tendencies which are each time becoming more orientated in the scheme of technology and explosive communications.

And if this prospective analysis today constitutes a powerful planning tool used by highly developed nations, which might contribute to deepening the tremendous economic, social, and political differences between them and Third World nations, it then appears logical that we in Venezuela, exceptionally endowed with such a powerful instrument for economic development as petroleum—but which is a temporary resource because it is perishable—should concentrate on long-term programming of the destiny that we must give to service contracts in relation to our national prospective.

This seems even more essential when we look back at the improvisation and lack of planification that has so frequently characterized the fifty-three years of our oil industry. Throughout that long length of time we have been unable to develop our own industry nor have we been able to control the destiny of our petroleum proper. We forget that several European nations, as well as Japan, were able to reconstruct not only an industry, but even a flourishing economy over a short span of fifteen to twenty years.

Due to a series of circumstances and lack of adequate planning, structuring a plan for service contracts was extensively delayed. Worse still, necessary measures were not taken for rapidly forming a legion of technicians, experts, and oil administrators, essential to giving real impulse to the Venezuelan Petroleum Corporation and the service contracts. Here we have another chapter wrought with inveterate lack of foresight in our petroleum policy, and because of its significance, it forms the nucleus of this final part of my statement.

I believe, Gentlemen of Congress, it has been sufficiently demonstrated that, notwithstanding the progress in our industry, we have been unable to appropriately control this activity despite its fifty-three years of existence, due to lack of foresight, preparation, and long-term planification. It is therefore absolutely essential for us to carefully analyze our medium- and long-term ideas in relation to this new instrument—service contracts—which we are about to introduce for the first time in this Parliament. These ideas do not seem yet to have been sufficiently clarified.

We obviously and undoubtedly need to reasonably expand our oil industry. Not because we arbitrarily feel like it, nor because we wish to enter a competitive race with other producing countries having reserves higher than our own; nor by further delivering our oil business to foreign companies; but rather because we now need, and will require for a long time, our petroleum income to build a nation; to educate an explosively growing population and to provide it with appropriate health and housing; to become independently industrialized, to develop a thriving agriculture; to export our products once we win the battle of productivity; and, in general, to construct a diversified country having adequate permanent resources for our future generations.

If we truly wish to build an economically independent and stable country, with institutions and installations that are definitive and efficient on all orders and levels of national life, framed in a pragmatic sense of nationalism, founded on the basis of today's available petroleum resources, we shall be unable to do so merely with words. We shall need, and with utmost urgency, to create a legion of experts and administrators for our oil business, permitting us thus to attain a true operative participation in this field, through service contracts and through rapid expansion, but avoiding administrative squanderings in the Venezuelan Petroleum Corporation.

Results Obtained through Service Contracts

After a three-year exploratory period, it is regrettable that drilling results hardly proved to be satisfactory. Out of twenty-one exploratory wells drilled, five were producers, seven were suspended, and nine were abandoned. Blocks B and D (Shell Sur del Lago, C.A. and Occidental Petroleum de Venezuela S.A., respectively) showed no oil deposits; blocks A and C (Occidental and Mobil Maracaibo C.A., respectively) proved to have only small deposits; and block E (Occidental) netted three producing wells.

INTERNATIONAL ASPECTS

Petroleum Supply and Demand

During 1964–1970, a quite tranquil oil era in Venezuela, its industry continued to decline. On an international plane, world demand for petroleum increased from 29 million barrels per day in 1964 to 45.8 million in 1970, of which the European Economic Community, the United States, and Japan absorbed more than 57 percent.

World production of crudes increased throughout this period at practically the same rate as demand, reaching an average annual growth rate of 8.5 percent (compared to 7.9 percent in consumption), increasing from 28 million barrels per day in 1964 to 45.7 million in 1970. In 1970 OPEC members supplied 86 percent of the crude entering international trading channels. That same year the Mideast countries contributed 30 percent of world crude production, reaching a daily average of 13.9 million barrels.

The production increase on the world level was maintained due to vast exploratory activities leading to the discovery of new reserves representing 309 billion barrels between 1964 and 1970, which, added to the existing reserves of 341 billion barrels, made 650 billion barrels. After production in those years was subtracted, remaining reserves were calculated at 543 billion barrels at December 31, 1970.

These figures clearly show the expansive tendency of the oil business worldwide and in the Middle East particularly. Intense explorations were made in other areas as well, such as the North Sea and Alaska. Thus, excluding Venezuela (and some tense areas such as Libya), the oil business continued to prosper throughout the world.

New Oil Sources

Increasing concern over finding new sources of oil supply, and increasing difficulties in finding them, obliged the oil industry to move further and further into more inhospitable areas. During the 1960s, the search for petroleum was intensified on continental platforms, in the open sea, and in the heart of the Arctic. Because of the results obtained, I shall specifically mention the North Sea, the North Slope of Alaska, the Arctic Islands (Canada), Siberia, and Colombia's and Ecuador's tropical jungles. Of these areas, those presently under exploitation (since 1970) are the Alaskan Arctic territories, the North Sea, and the deposits discovered in the Colombian and Ecuadorian jungles.

The North Sea has been subjected to intense exploration, leading to the discovery of important gas deposits. However, it was only in 1970 that the first great oil field in Norway's territorial waters was discovered; similar discoveries were subsequently made in the territorial waters of Great Britain, Holland, and Denmark. Perspectives in the North Sea Basin have improved day by day from the standpoint of potential petroleum reserves; discoveries of oil, condensate, and gas have multiplied as the zone is more intensely explored. However, soaring costs due to the inhospitality of the area and to technical and logistic problems pending solution make it difficult to predict the future of crude production in the North Sea.

Colombia's oil production lies in the jungles close to its frontier with Ecuador, and a pipeline was laid through the Andes for transporting crude to the Pacific coast. Colombian oil is used in its internal market, which is still partially dependent on imports to meeet its needs.

The deposits of Ecuador are located in the eastern zone of the country, south of Colombia's fields, and a pipeline was built—also trans-Andean— for transporting crude to the port of Esmeraldas. Ecuadorian production began entering the international market in August 1972.

The Soviet Union has faced technical problems in exploiting deposits in Siberia and has requested cooperation from Japanese and United States companies (such as Occidental Petroleum Company).

In Alaska, environmental save-nature committees managed to obtain a postponement of the construction of pipelines necessary for transporting crudes from the North Slope. It is for this reason that development programs were paralyzed until mid-1974.

Therefore, efforts made in exploring new areas have not been sufficient to meet the energy crisis affecting the world today.

ARPEL

From June 26 to 29, 1961, and on the invitation of the Venezuelan Petroleum Corporation, a meeting was held in Maracay, Venezuela, attended by state oil company representatives from Argentina, Bolivia, Brazil, Colombia, Mexico, Peru, and Uruguay, for the purpose of discussing the need to create an organization to unify them and promote agreements of common interest. After several sessions these countries, with the inclusion of Chile, held another meeting in September 1966, in Rio de Janeiro, Brazil, for the final constitution of the Latin American State Association for Reciprocal Oil Assistance (ARPEL).[10] All these countries are at present members of ARPEL, except Colombia, which withdrew, but which in November 1973 requested readmission as an active member.

The objectives of ARPEL are to study and recommend formulas of mutual cooperation for the defense of common interests and to exchange state oil company experiences in assuming a direct role in all phases of the oil business in Latin America and particularly in the respective internal markets. Among the subjects under study in 1974 and 1975 are commercialization in external markets and exploitation and transportation of heavy crudes.

OAPEC

On January 9, 1968, the Petroleum Ministers of Saudi Arabia, Kuwait, and Libya met in Beirut, Lebanon, and signed an agreement creating the Organization of Arab Petroleum Exporting Countries (OAPEC).

At present ten Arab countries are members of this Organization, since the founders (Saudi Arabia, Kuwait, and Libya) were joined by Abu Dhabi, Algeria, Bahrain, Dubai,[11] Qatar, Iraq, Egypt, and Syria. Of these,

---

[10]Integrated by the following state companies: Administración Nacional de Combustibles, Alcohol y Portland (ANCAP), Uruguay; Corporación Venezolana del Petróleo (CVP), Venezuela; Empresa Colombiana de Petróleos (ECOPETROL), Colombia; Empresa Nacional de Petróleos (ENAP), Chile; Petróleos Brasileiros S.A. (PETROBRAS), Brazil; Petróleos Mexicanos (PEMEX), Mexico; Petróleos del Perú (PETROPERU), Peru; Yacimientos Petrolíferos Fiscales (YPF), Argentina; Yacimientos Petrolíferos Fiscales Bolivianos (YPFB), Bolivia.

[11]Abu Dhabi and Dubai belong to the United Arab Emirates.

seven are members of OPEC (Algeria, Iraq, Kuwait, Libya, Qatar, Saudi Arabia, and the United Arab Emirates). Originally, membership in OAPEC was restricted to Arab oil-exporting countries which were also members of OPEC; however, due to statutory reforms it is now sufficient to be an Arab oil-producing state, whether an exporter or not, and hence not necessarily a member of OPEC.

Through its name proper, it would appear in principle that OAPEC is an organization of Arab world countries parallel, and similar, to OPEC. Most of its promoters were members of OPEC, and the purpose of unity was their common condition as oil exporters. However, there are certain differences between the two organizations that might be summarized as follows:

OPEC unites governments of export countries for the main purpose of regulating and controlling prices and production.

OAPEC unites oil-producing Arab states, exporters or not, for the purpose of protecting the interests of Arab oil and for directly participating in the petroleum industry by means of joint investments through international consortia integrated by themselves.

OAPEC's main objectives can be summarized in the following three points, described by Sheikh Ahmad Zaki al-Yamani:

1. That petroleum activity become a tool for protecting the interests of member countries and of the Arab people in general.

2. That opportunities be created permitting member states to make joint investments in the oil business.

3. That direct relations be established with oil-consuming nations with a view to expanding markets.

From these objectives, it can therefore be assumed that OAPEC is principally an organization grouping oil-producing Arab states for the purpose of uniting them in their common interests, i.e., political, ethnic, and economic.

In the economic aspect, one fundamental characteristic is that OAPEC pursues the cooperation of its members in direct participation in the petroleum industry in its various phases of exploration, exploitation, refining, and transportation, with a view to realizing joint investments in the oil business and to forming international consortia. Within this objective, I shall mention two important projects of OAPEC member states.

One project is the construction of a dry dock in Bahrain (scheduled to be placed in service in 1975). By June 1974, the agreement for the construction of this project had been ratified by five countries: Bahrain, Iraq, Kuwait, Libya, and Saudi Arabia.

The other project consists in the creation of an oil fleet owned by OAPEC member countries that will operate as a multinational company under the flags of each and every one of the affiliated states. The capital foreseen for this project is $500 million, raised through joint investment. At present the first four tankers of this fleet are being built: two in France of 278,000 deadweight tons each, and two in Germany of 318,000 deadweight tons each. These units will be delivered between December 1976 and April 1978.

OAPEC's significance in the world petroleum industry is substantial, not only because of its economic objectives, but also because the world's main producing and exporting countries are concentrated therein.

The Shultz Report and the United States Policy

On February 20, 1969, Richard Nixon, newly elected as President of the United States, named a Cabinet task force for the purpose of formulating recommendations covering one of the most thorny, complex, and difficult subjects that his administration had in such a short time to confront: a policy for oil imports.

The task force was headed by George P. Shultz, then Secretary of Labor; the other members were the Secretaries of State, Interior, and Defense, the Director of the Office of Emergency Preparedness, and Robert Ellsworth, a White House advisor. At its first meeting (April 8, 1969) the task force appointed Phillip Areeda as its executive director, a professor of economics and law at Harvard, who also headed the staff committee. Areeda, a specialist in antimonopolistic legislation, had been a consultant to President Eisenhower and had participated in deliberations leading to the mandatory control over imports in 1969.

On January 21, 1970, the Caracas newspaper *El Nacional* published the text of the working document, which gave rise to several different reactions. The Minister of Mines, although not making any fundamental comments, declared that the Shultz report could prove to be favorable for Venezuela's oil industry. The government's first declarations were hence optimistic. However, according to press reports, this initial optimism seemed to wither later on in governmental spheres; it was even rumored that the Minister of Mines and Hydrocarbons was going to Washington for the purpose of transmitting, at top level, the Venezuelan government's technical and political observations, even though an eventual meeting between Presidents Caldera and Nixon was being contemplated. It would appear that the government gradually had become aware of, and studied more thoroughly, the recommendations contained in the report, raising many afterthoughts and preoccupations.

Actually, an analysis of the working report leads to the conclusion that it is systematically and shrewdly conceived, apparently contemplating

United States interests only and disregarding those of other countries, in particular Venezuela, its traditional, safe, and consistent petroleum supplier.

Shultz Task Force Recommendations

Because of its fame at the time, and to illustrate the radical change in market conditions, and hence the world focus on oil problems, over a short five-year period, I shall briefly summarize the recommendations of the Shultz report.

1. To obtain United States consumer savings through a system of differential tariffs that will permit a limited flow to the United States of cheaper crudes proceeding from Latin America and the Eastern Hemisphere (Middle East and Africa), which, on competing with the United States internal market, will cause a drop in general price levels of domestic and imported crudes.

It was estimated that this saving by United States consumers would represent more than $2.5 billion in 1970 and $4 billion in 1980. However, it is now obvious that, on account of the critical oil supply situation facing the United States, forcing rationing in certain regions, it was even absurd to assume that a flow of cheaper crudes proceeding from both hemispheres could be channeled to the United States.

2. To eliminate the ever-increasing risk of corruption arising from the system of quota assignments, which has engendered so many controversies.

Undoubtedly, the system of assigning import quotas to United States refineries opened the door to corruption and favoritism, and it is interesting to see that the United States government and public opinion were conscious of this problem. On the other hand, such quotas caused a transfer of Venezuela's income to refineries in the United States, with no benefit whatsoever to Venezuela. The Shultz report thereby constituted a testimony to the corruption generated by the quota-assignment system, which had been imposed by the United States government through so many years.

3. To apply a concept of *national security* to ensure normal oil supplies in case of emergency, and in general, to prevent detriments to the uninterrupted future development of the United States economy. The following is a fundamental aspect of this concept contained in the report:

The security concept from the viewpoint of oil supplies proceeding from other countries   The report basically stated the need for the United States to be able to rely on a regular flow of imported oil in order to meet its total hydrocarbon requirements (especially crude), which must not be subject to any interruptions under any circumstances whatsoever. It was essential to avoid major dependence on countries whose shipments could be cut off or drastically cut down by threat of war or for any other reason (as in fact occurred in the Middle East in the critical period between November 1973 and March 1974).

To avoid such a risk—according to the report—the United States should first be able to increasingly rely on oil supplies from Canada. Hence the recommendation to exclude Canadian imports from the proposed tariff system and the evident open proposal for a United States–Canada energy and oil integration. In the second place, the United States would be able to rely on supplies proceeding from Latin America. The report obviously referred fundamentally to Venezuela, as throughout its context it directly mentions Venezuela. In the third place, the report considered supplies proceeding from the Eastern Hemisphere as being very unsafe. Therefore, the differential tariffs for Latin American countries proposed in the report are slightly lower than those recommended for crude proceeding from the Eastern Hemisphere.

Events were later to prove that the report was overoptimistic concerning future imports that the United States could expect from Canada. In fact, the Canadian government establishes its export volume of crude by taking into consideration its own domestic consumption and its self-supplying capacity over a certain length of time. Consequently, in the face of abnormal circumstances, Canada was bound to suspend or restrict its sales of crude to the United States, as it indeed did during the energy crisis of 1973, when it decreased exports to that nation by 49,500 barrels per day.

4. *Integration with Canada*. The report points out that the risk of political instability or animosity is by far more unlikely in Canada, compared to other petroleum countries.[12] Furthermore—according to the report—it was improbable that Canadian oil would be exported to markets other than the United States, since nearly all exports were made via the terrestrial Great Lakes Pipeline. Consequently, the report proposed a definite energy integration with Canada, implying:

   a. An agreement for the construction of a trans-Canadian pipeline on an equitative cost basis, and recognition by Canada of unobstructed transit rights.
   b. The joint development of the Alberta asphaltic sands, for which purpose Canada's provinces would eliminate licensing restrictions so as to open up the market to the United States. Both governments were to facilitate this development by means of mutual agreements covering investigation and tax policies, in order to accelerate production of the synthetic oil resources of both nations.
   c. An obviously negative point for Venezuela was the observation

---

[12]Where this risk was concerned, it would appear as if Venezuela were being placed alongside all other countries—including those of the Middle East and Africa.

that "it was difficult to justify the United States' full tariff preference for Canada as long as Canada continued importing its East Coast requirements from sources beyond its borders, especially because, in an emergency, Canada could decrease its United States supplies by interrupting or decreasing its own imports of crude from other countries." In the development of this policy of integration, the report contemplated the advisability of a "common external tariff," or that Canada should seek "other means for limiting its dependency on imported oil to a proportion more concordant with its rate of consumption."

The importance given by the report to this energy integration between the United States and Canada is shown by the explicit recommendation that, in order to avoid problems with other export countries, it would be advisable that all these measures be included within the context of a formal and structured energy integration agreement, thus diminishing "the intrinsic value of the system of preference to Canada, and at the same time making it *more acceptable from the point of view of diplomacy.*"

We must remember that the visible deterioration of Venezuelan exports to the United States market in the 1960s—added to the prevailing import-quota policy and, especially, to the discriminatory treatment in favor of Canada, enabling it to boost its oil sales to the United States at Venezuela's expense—was compensated to a certain degree by the important increase in Venezuela's crude exports to the east coast of Canada.

The integrated oil policy for all of North America has been a subject of consideration for several years. The *Petroleum* Press Service in February 1969 stated that "the true wish—not to say the only hope—for a flourishing growth of Canada's oil industry during the forthcoming years depends upon the prompt creation of a continental oil policy for North America." The article also indicated that "Canadian reserves were increasing twenty-two times more than its current production, whereas those of the United States had declined, reaching the alarming proportion of ten to one."

    5. *The Differential Tariffs*

    a. Concerning Canada. The report foresaw that, should all quantitive restrictions covering entry into the United States of Canadian crude be eliminated, by 1970 Canada would be exporting 3 million barrels per day to the United States; part would go to the East Coast through the projected trans-Canadian pipeline, which would transport oil from the Arctic (of Alaska and of Canada) plus crude produced in Alberta (Canada).

None of these predictions ever took place. In 1970 Canadian exports to the United States had reached 672,000 barrels per day instead of 3 million; and in order to secure its domestic needs, in February 1973

Canada decided to reduce its exports to the United States. Furthermore, the pipeline that was to join the fields of the North Slope of Alaska with those of Alberta, and later distribute crude by using existing pipelines in the Great Lakes region and in eastern Canada and the United States, has not yet gone beyond the planning stage.

   b. Concerning Latin America (Venezuela). The report indicated that, although there were reasons that favored granting Latin America the same unlimited access to the United States market as explicitly recommended for Canada, such treatment would be incompatible with other aims of the program. On the other hand, it pointed out that, in contrast to Canada, a higher price for Venezuelan crude would not cause an increase in Venezuela's production, and that by selling Venezuela's production at world prices (or at an approximated figure), any preferential treatment would create fortuitous profits for oil companies operating in Venezuela. Insofar as service contracts were concerned, according to the report, the Venezuelan government—and the companies—would use such a preference as a basis for setting the terms thereof, and possibly in such a case the final beneficiary would be the Venezuelan nation.

   6. *Fuel oil.* I have pointed out some aspects concerning which the Shultz report is clear and definite in its recommendations. For example, the concept covering United States national security; the preferential treatment for Canada; and the energy integration agreement with Canada. The treatment of fuel oil is a similar case, regarding which the report concretely suggests and recommends (presumably clearly protecting Venezuelan exports to the United States market) the continuation of the prevailing system—which does not, essentially, establish any control over fuel oil insofar as District I imports are concerned; furthermore, it recommends that this program be extended to other districts.

   It is also explicit in pointing out the desirability of desulfurization to be accomplished within the United States, taking into account preferences granted to Latin America (I interpret this in the sense that desulfurization plants already existed in Venezuela) and leaving inherent the restrictions on imports of fuel oils proceeding from the Eastern Hemisphere.

   On the other hand, the report points out that even if Venezuela and other Latin American producers guaranteed not to divert their hydrocarbons toward other import markets in a case of emergency and without consent of the United States government, the present dependency of Europe and Japan on relatively insecure Eastern Hemisphere sources is such that in a critical emergency the United States might be obliged to accept such diversion of supplies, and Latin American exports to the United States would decrease proportionately.

   Undoubtedly such a situation could quite easily arise, but it also could apply to Canada. Due to its multiple ancestral ties with France and Great Britain, Canada might also be obliged in an emergency faced by those countries, as well as by other European nations, to divert some of its United States exports to them. The oft-commented-upon reduction in Canada's petroleum supply to its southern neighbor has already become symptomatic in this time of the most critical energy crisis faced by the United States in recent years.

Finally, there is another statement by the task force worth mentioning, which reflects the attitude traditionally adopted by United States governments in "taking Venezuela's oil for granted." I refer here to the report's conclusion that "a selective boycott against the United States by Eastern Hemisphere producers could be counteracted through imports proceeding from Venezuela." Thus, Venezuela would be a kind of reserve, a "standby," in drastic cases, such as a boycott by Middle East or African export countries against the United States. This has been, in part, recent history, and is once again repeated in the Shultz report.

It is interesting to note that, as with Canada, the Shultz report was too optimistic concerning Venezuela's production capacity—which, as I have mentioned, was in those days, and still is today, facing a process of decline. It appears that it would have been by far more constructive, and obviously more favorable to United States interests, for the report to have studied and suggested—with the same thoroughness it dedicated to analyzing the possibilities for an energy integration with Canada—some kind of plan for reactivating the petroleum industry of Venezuela, the main oil supplier of the United States.

In the light of events occurring in the oil world, in particular since the end of 1970, we can see that the Shultz report was excessively optimistic in many of its conclusions. However, we also know that when the report was drawn up, prevailing and foreseeable short-term circumstances did not easily indicate the possibility of any radical change in the world oil panorama, nor was it easy to predict any emergent crisis in energy supplies. There was talk of a deficit in oil supplies, but only in moderate quantities, since the regular increase in demand did not appear to contain any surprising elements; nor was it possible to foresee the decisive measures adopted by oil-producing countries in relation to prices and to their quest for a higher control over the oil industry. Neither was it easy to foresee delays in other projects destined to set into motion other sources of energy, such as the Alaskan deposits or nuclear generators. However, a conglomeration of several unforeseen events, emerging simultaneously, was to override whatever forecasts were made at that time.

To enter into a deeper analysis of the failures that we can today detect in the report submitted by the Shultz task force in 1970 would have no value other than merely as a historical analysis. Events that have occurred are actual facts, and we are able to use them as evidence in our critical judgements. However, at the margin of these errors two points predominate: first, the elimination of import quotas and their substitution by a more flexible system; and second, the possible energy integration with Canada, a project that is still alive, although it has not yet been successfully achieved, despite United States diplomatic approaches to Canada.

EIGHT

Economic Dimensions of the Venezuelan Oil Industry's Decline

I believe it would be advantageous to speak of the opinion of men deluded by their self-interest who should be enlightened by their true interest.

SIMÓN BOLÍVAR
Letter to General Francisco de Paula Santander,
El Rosario, Cúcuta, May 30, 1820.

Venezuela is a country encumbered by innumerable problems, and a multiplying element at their roots is the explosive growth in population. At an approximate 3.4 percent average annual rate, Venezuela's population is increasing twofold every twenty years. This factor is, of course, accompanied by a spiraling growth in a series of needs that must be met jointly by the state and by the nation.

As stated by the president of Venezuela's Social Security Institute, among the five principal causes of death in Venezuela are undernourishment and avitaminosis, i.e., hunger. This appears to be incredible in a country so frequently classified as "rich," which undoubtedly it is if we compare it with most others in the process of development. However, added to the many problems arising from its demographic growth, Venezuela faces the no less alarming problem of a qualitative nature, namely, the gradual deterioration of vast undernourished and impoverished population sectors, having a nonexistent or very low level of education and the most intensive rate of natural increase close to 5 percent per annum.

The country is facing serious difficulties in relation to unemployment, overcrowding, environmental pollution of its main urban centers, and education. The agricultural sector is in a nearly chronic state of crisis, and industrial development, although in much better shape, still needs a more vigorous impetus in order to meet its great potentialities.

To enumerate the grave problems afflicting Venezuela would be by far too extensive and furthermore beyond the scope of this book. However, I feel it is essential to point out the marked inefficiency that is continually in evidence at all levels of public administration. A consequence thereof is the tremendous squandering of funds that rightly belong to the entire nation and that are overwhelmingly derived from the gradual liquidation of the country's petroleum assets.

How has the state been dealing with such problems? Through increased public expenditure, nourished in an extremely high proportion—as we shall see—by oil. Total state fiscal expenditures cannot, technically, be measured by annual budgets; they should be accurately measured by consolidating all public expenditure. In the last twenty years especially, the phenomenon of decentralization of public expenditure has been occurring, creating numerous payouts that do not appear in annual budgets.

Public autonomous institutes, or state companies, have, in addition to budgeted funds, their own income, thus further increasing the total (consolidated) expenditure. Many of these institutes, such as state companies, generate income (transformed in its majority into expenditures) through the sale of assets and contracting of loans, which neither appear in budgetary statistics nor are submitted to the general comptrollership of the nation or to any other official monitoring.

It is thus that measurement of these official payouts should be made through the consolidated public expenditure, which comprises the total disbursements effected by the central government, the regional governments (20 states and 2 federal territories), the 147 municipalities, and the 70 para-state organisms represented by 37 service entities and 33 companies within the public sector. This statistic has not been made available on a regular basis.

Let us briefly review the evolution of the national budget of expenditures in recent years and the budget approved for 1975. Between 1960 and 1970, Venezuelan total fiscal income increased at a geometric average rate of 6.7 percent, on rising from Bs4.9 billion ($1.6 billion) in 1960 to Bs9.2 billion ($2.2 billion) in 1970. Furthermore, this growth in fiscal income maintained practically the same tendency even up to 1972, when it reached a level of Bs16.0 billion ($3.8 billion).

The effects of the series of measures adopted by Venezuela in this decade in relation to oil prices and taxes were felt since 1973—especially as a consequence of the increase in the export values of Venezuelan crude—and they radically changed the country's fiscal income. The Venezuelan government had estimated its normal income for 1973 at Bs14 billion ($3.3 billion), whereas in actual fact it received Bs16 billion ($3.8 billion); the 1974 budget for ordinary income predicted and income of Bs14.4 billion ($3.4 billion), whereas the amount collected reached approximately Bs42.5 billion ($10.1 billion). Insofar as 1975 is concerned, it expects to receive Bs41.5 billion ($9.8 billion), according to the budget law approved by Congress. This 1975 budget—although slightly inferior to income received during 1974—is nearly three times higher than the $3.4 billon budget approved for 1974.

We can thus clearly see the progressive increase in state income—and hence expenditures—which is a direct consequence of increased oil income. Petroleum income was 60 percent of Venezuela's ordinary income in 1960; in 1970 it was 61 percent, and a figure of up to 86 percent has been estimated for 1975.

An analysis of the composition of expenditures, especially covering investment and current expenditures, is beyond the scope of this book. Neither do I have at my disposal the space necessary for a study of the consolidated public expenditure, which is by far higher than forecasted by the national budget, since it includes para-state organisms (public service entities and companies referred to previously) that have exorbitantly multiplied and grown in recent years. This, in turn, has originated the explosion in bureaucratic expenditure and the consequential squander of funds that proceed, mostly, from Venezuela's nonrenewable resource, petroleum.

We must also bear in mind that the payouts that still appear as investments in national budgets and in the budgets of public entities and companies in practice—in major proportion—feed the machinery of that increasing bureaucracy.

If Venezuelans of today wish to legate to their future generations the prosperous, educated, and stable Venezuela that should logically be the fruit of this prodigious petroleum era, then this grave situation should be submitted to immediate and urgent corrective measures. In regard to this situation, my observations are the following:

1. Throughout the last two decades a giant has been growing within the Venezuelan economy: the state enterprise. The Venezuelan state, besides attempting to satisfy traditional public needs (education, health, security, road systems, etc.), has created a tremendous complexity of enterprises, emerging as an impresario within the siderurgical, banking, sugar, hotel, shipping, shipbuilding, airline, marketing, and other fields.

2. This state-impresario colossus tends to expand rapidly not only

through the vertiginous growth of its budgets, but also through the continuous creation of new organisms.

Without making any precise judgment as to in which cases the state should appropriately satisfy public needs, which it is endeavoring to meet through new state institutes or companies, it is evident that the phenomenon of their creation or expansion, as well as the increase in their payouts, could in itself create critical fiscal situations of a functional nature, over the medium and long term. This is especially true in view of the further deterioration of Venezuela's traditionally inefficient public administration and the lack of adequate controls.

3. Alongside this colossus exists another, from which the state has been deriving an extremely high proportion of its income: the oil industry. Its properties, 90 percent of which are still held by the private sector, will shortly revert to the state, once the anticipated reversion takes place. The nationalization of the iron industry, which took place on January 1, 1975, should also be mentioned. All this means:

   a. That a new expansion will occur—far more intensive than in the past—in the area of public administration, with the consequential risk (which should be avoided because of its obvious grave dangers) that substantial incomes proceeding from such nonrenewable resources might be squandered through excessive bureaucracy.

   b. That large investments, which up to now have been made by the concessionaires, must gradually be provided by the Venezuelan state, if it is to maintain an appropriate growth rate in its oil industry once direct operation is assumed. These new investments might very well amount to billions of dollars over a very short span of years.

4. Added to the two preceding giants, there is a sector that is comparatively small: Venezuela's private economy, which generates less than one-third of the state's income.

I must not fail to mention that the two industrial sectors that are in the process of becoming administrated directly by the state—iron and petroleum—will generate a series of activities, both industrial and in respect to services, that will be provided by the private sector.

Where iron is concerned, nationalization having occurred on January 1, 1975, the Venezuelan government has established a policy whereunder private investors and impresarios (Venezuelan and foreign) will participate in the various processing phases and in the production of steel products. Likewise, their association with the Venezuelan state in mixed companies is foreseen.

Insofar as petroleum is concerned, nothing has yet been defined as to participation by the private sector in the industry's related activities, because nationalization has still to take place. However, I believe there will be a wide field opening up to private entrepreneurs, essentially in the varied service area required by petroleum activities.

5. Notwithstanding the unprecedented growth in total public expenditure, the state is obviously still not adequately satisfying the people's needs, as is evident by the intense poverty and lack of opportunity still prevailing in dense sectors of the population.

Regardless of the foregoing factors, the nation has somehow managed to survive an accumulation of threatening problems, and its democratic system has become assured because the state has been able to count on abundant and increasing returns from the oil industry. Let us now take a look at Venezuela's dependency on oil.

OIL DEPENDENCY

All Venezuelans are aware, to a greater or lesser degree according to their knowledge and information, that Venezuela's economy depends very highly on its oil industry. Let us try to measure the current situation in order to determine if this dependency has decreased in recent years, maintained the same intensity, or become more pronounced.

It may be said in a general sense that Venezuela's oil income has continuously increased. From $1 billion in 1959, it increased to $1.7 billion in 1971; furthermore, it will continue increasing as a result of higher sales prices in world markets and higher export reference or value prices used in income tax assessment.

Under these circumstances, undoubtedly the dependency of Venezuela's national budget on oil income will become even more pronounced in forthcoming years, unless the returns are considered (as they were recently) as extraordinary income instead of ordinary, since they proceed from the liquidation of a national asset and hence are in actual fact extraordinary.

Participation of petroleum in the formation of the gross national product tended to climb until 1957, when it reached 31 percent, but it then began to decline, falling to 29 percent in 1973. This would be a positive factor in the Venezuelan economy if it were due to a greater growth in the rest of the national economy. However, the decreased participation of oil in the gross national product is due rather to the gradual slowdown that the industry underwent during the 1960s.

The percentage of foreign exchange generated by petroleum in Venezuela is overwhelming. In 1958 this percentage reached its most serious level: of $1.30 billion received by the Central Bank of Venezuela, $1.29 billion proceeded from the oil sector, i.e., 99 percent. In subsequent years the percentage fluctuated between 62 percent (1960) and 91 percent (1973).

Another dominant aspect of the oil industry's role in the Venezuelan economy is reflected by the gross total of foreign investment. In 1960, 61 percent of foreign capital corresponded to oil companies. During the 1960s oil investment decreased, and in 1970 it stood at 50 percent, still excessively high for any country. However, the decrease was achieved not through any increase in other activities, but rather through a process of investment withdrawal, which took place in the nation's oil industry in the 1960s. Another reason for the decrease is the fact that since 1971 official statistics have segregated refinery investments from the oil sector and included them in the industrial and commercial sector. This change was presumably made in order to make the oil industry's proportion appear less preponderous in total foreign investments.

From the preceding it can be seen that Venezuela depends on petroleum in a proportion dangerously high for its economy. Furthermore, its oil industry, 95 percent tied to foreign markets, is increasingly concentrating its exports toward one single market: the United States. Imports by the United States (including Puerto Rico), in 1960, 1971, and 1973, reached 48 percent, 55 percent, and 61 percent, respectively, of Venezuela's total oil exports. If we add to these figures oil imported by Canada from Venezuela, we can clearly see how Venezuela is increasingly dependent upon the North American market. It seems paradoxical that while Venezuela is keenly endeavoring to take over the control of its oil industry, it is simultaneously concentrating its exports toward the North American market.

THE DECLINE OF VENEZUELA'S OIL INDUSTRY

Let us now examine the decline that has been taking place in Venezuela's oil industry since 1958—especially during the 1960s—using such figures and indexes that might objectively show us whether this decline did, in fact, originate during the stage under analysis (1958–1975). I am not taking into account any intermediate stationary situations because, in general, when industries are not expanding (especially the oil industry, which requires such large investments), it is difficult for them to remain stationary over a long period of time. The tendency finally is to decline.

Expansion or decline in petroleum industries depends fundamentally on hydrocarbon reserves and other assets on which capacity for production, refining, and other pertinent activities is based. We must in this sense observe that the increase in profits obtained by oil-exploiting companies, taxes paid to the state, and dividends distributed to stockholders, although possibly giving an impression of prosperity, do not always correspond to a status of development in the industry proper. Such factors might depend, principally and totally, on special circumstances such as price increases, increased tax rates, and even internal company decisions to distribute larger dividends to shareholders.

Quite to the contrary, these apparently positive external factors might lead to negative results. For example, profit sharing in high proportion might decrease investment, which in turn could contribute to the decline of the industry proper. As far as greater fiscal participation is concerned, this is obviously convenient for the nation, especially if the increase in taxable income is due to a higher participation per barrel rather than to increased production. However, increase in state income, especially in this case, does not necessarily constitute an index of expansion in the industry.

During the 1960s, Venezuela's higher fiscal participation was indeed due to an increase in income per barrel rather than to production increase, the rate of which was very moderate. This, in itself, represented another positive aspect, because the nation received more per relative quantity of a nonrenewable product. However, this aspect—apparently favorable—did not help to clarify the fact that Venezuela's oil industry proper was being jeopardized; instead, it created a climate of bonanza in the country that in a certain way even served to cloak the truth of its declining oil industry.

In fact, fiscal participation, which was $0.81 per barrel in 1960, increased to $1.03 per barrel in 1970. In subsequent years fiscal profits per barrel continued increasing progressively, spiraling to extraordinarily high figures: $1.30, $1.60, and $2.33 in 1971, 1972, and 1973, respectively, and close to $9.00 estimated for 1974. Nevertheless, since 1970 oil production has increased only moderately, with a general tendency toward decline. In the first ten months of 1974 production decreased tremendously: nearly 11 percent under its 1973 average, whereas fiscal income reached its highest peak ever.

This pronounced shrinkage in Venezuela's oil production is partly a consequence of decreased consumption, especially by its main customer, the United States, which absorbs approximately 40 percent of exports. It is also due to the decision by OPEC countries to reduce production in order to avoid flooding the market and possibly weakening price levels. But, it is the consequence as well of the reduced capacity of Venezuela's oil industry, caused by its decline that began at the end of the 1950s, as we shall soon see.

The following paragraphs are transcribed from the 1969 Memoir of Venezuela's Ministry of Mines and Hydrocarbons and certainly lead one to ponder deeply:

> Since 1965 a decrease in reserves has become evident as a result of production's being higher than the proportion of additional discoveries, expansions, and recoverable reserves. From 1961, discoveries and expansions remained at very low and decreasing levels; only recoverable reserves increased, so that between 1961–1964 there was little negative change in reserves. In general, should the present situation persist, reserves will continue decreasing at a rate complementary to the rate of oil production.
>
> This situation is aggravated by the fact that light and medium oils, i.e., those having a gravity higher than 25° API, are extracted at a rate doubling that of heavy oils, even though their volume of reserves is similar.
>
> It is on account of the foregoing that the Technical Hydrocarbons Office wishes to expand its personnel teams in order to truly and rapidly resolve these problems.
>
> However, this is not a simple task because Venezuela is riddled by a series of circumstances difficult to overcome: recruitment of capable personnel; payment of adequate salaries; expansion of offices; the need to make systematic use of computers, to indoctrinate personnel, etc. Hence, if the people involved, by reason of hierarchy or knowledge, do not pay immediate and thorough attention to these problems, satisfactory solutions will be further delayed and possibly when such solutions are found *it might very well be too late.* [1]

Another important symptom of Venezuela's oil decline is the systematic layoff of workers that has been taking place in the industry. White- and blue-collar employees and laborers working in the oil industry in 1958 totaled 44,720. This number has decreased year by year, dropping to 22,942 in 1973.

New techniques and automation have considerably increased productivity in the world oil industry during recent times, creating a tendency toward decreased employment. However, the magnitude of personnel reduction in Venezuela's oil industry, to almost half its total in the period under study, is a characteristic not of a prosperous expanding industry but rather of an industry in frank decline.

Whereas in 1958 production was 21,265 barrels per year per worker, in 1972 this figure more than doubled, to 52,182. If we consider that total production increased from 951 million barrels in 1958 to only 1.3 billion in 1972, i.e., by 36.1 percent, we may conclude that the increase in productivity (number of barrels per worker) was principally achieved on the basis of a drastic reduction in the number of workers and not through

[1]Ministry of Mines and Hydrocarbons Memoir, Caracas, 1969, pp. VIII–2 and 3. The italics are the author's.

increased investment (rather, investment in net fixed assets decreased by nearly $230 million).

Massive layoffs of laborers and other employees occurred during this period, notwithstanding employment stability clauses contained in collective contracts. Concessionaires managed to do this in many cases by offering special payoffs that induced the workers to accept layoffs.

During this 1958–1972 period, and notwithstanding noninvestment in the oil industry and the grave situation relating to reserves, although the average production rate of increase did not reach 2.5 percent annually (in 1972 it decreased by more than 9 percent in relation to 1971), the state's income did increase considerably. In addition to the moderate production increase mentioned previously, this was due to increased participation (the 60 to 40 percent 1960 relation became 89 to 11 percent in 1972) and to price increases since 1970.

FINAL COMMENTS

From the foregoing it is clear that Venezuela is in a paradoxical position: It cannot fully enjoy the unprecedented advantages offered by current opportunities on the international market, notwithstanding the extensive potential reserves that it apparently still possesses. It is therefore of immediate urgency for Venezuela to become fully aware of this situation, at least as a first step toward adopting measures leading to its rightful path. Venezuela should especially bear in mind that the favorable phenomenon of such an extraordinary oil-price increase, although originating much higher profits per barrel to the nation, contributes further to hiding the process of decline occuring in its oil industry.

I believe it is of utmost urgency that appropriate decisions be taken in order to ensure that the following take place:

1. Investments should be made on a large scale and as soon as possible, especially where exploration is concerned, in order to reinforce Venezuela's production capacity. Such investments should be made in:

   a. Existing concessions

   b. New areas (of conventional and nonconventional oil)

Similarly, tremendous efforts should be made to increase, as far as possible, the recovery factor (secondary recovery).

In the light of Venezuela's present situation, there would be two alternatives:

   a. Development would be undertaken exclusively by the state.

Applicable measures should eliminate traditional state inefficiency, which could represent loss of huge reserves belonging to all Venezuelans;

on the other hand, the state should urgently invest in the formation of qualified personnel.

    b. Development would be undertaken by the state in partnership with private Venezuelan and foreign capital.

A great flexibility exists that would permit the combination and instrumentation of public and private associations. The Soviet Union's example is very illustrative in this respect.

In Chapter 10 of this book I shall refer to these points in further detail.

    2. Decisions should be taken toward increasing, over the short term, the capacity of desulfurizing fuel oils. With the Amuay refinery expansion carried out by Creole Petroleum Corporation completed at the end of 1972, Venezuela has a desulfurizing capacity of only approximately 400,-000 barrels per day. If this capacity is not increased, Venezuela, with exports of fuel oils at approximately one-third of its total, might witness a shrinkage in a vital market, the market for fuel oil of low sulfur content.

It is to be remembered that Venezuela must also increase its production proportion of heavy and medium-heavy crudes.

    3. Finally, Venezuela should take the following very much into account:

    a. Decisions related to most large industries, especially in such a heavy and complex industry as that of oil, take years to produce results. It could very well happen that satisfactory solutions, when taken, might come too late. If this latter should occur toward the end of the present decade, a pronounced decrease in Venezuela's oil production could take place, and it would be attributable not, as at present, to marketing and price considerations, but rather to the decline of Venezuela's productive capacity.

I feel that the Venezuelan government should do more than shape a policy wherein fundamental aspects are taken into account (such as those mentioned previously, and also at the end of this book). The government should also immediately proceed to elaborate a program of production rationalization, relating to the volume and composition of petroleum reserves, in order to maximize preservation of future productive capacity and to avoid, wherever possible, any abrupt drop therein during the second half of this decade. This policy has been partly adopted, but for another reason: in order to contribute to the preservation of the current level of world export prices.

This production program, with the additional object to implement appropriate conservation measures, and already initiated by Venezuela's present government, might imply fiscal sacrifices by creating a systematic reduction in production in forthcoming years while attempting to revitalize and expand the industry through the incorporation of new reserves.

These new reserves would permit an increase—or at least a stabilization— of Venezuela's productive flow, as a consequence of intense activity in exploration and secondary recovery, added to a vast development of its natural gas industry.

Alongside the preceding factors, which are pressing toward a highest possible fiscal return, it should not be forgotten that within a few years (possibly in the 1980s), investments being made by great industrial nations for the purpose of developing their own alternative sources of energy (nuclear, geothermic, solar, nonconventional petroleum, etc.) might start having effect. Furthermore, the speed at which these alternatives are developed quite possibly will be intensified because of the alarm engulfing these nations over their acute dependency on oil-producing countries, especially of the Middle East.

Should this occur, then possibly oil-price increases will be limited. Furthermore, within the present panorama of the world's energy crisis, should oil prices spiral to a vertiginous level, other fuels, such as desulfurized carbon, might become competitive, and hence another limitation to the rapid increase of oil prices might appear.

   b. Venezuela's position in any OPEC agreement is also directly proportionate to Venezuela's productive capacity.

   c. As long as Venezuela increasingly depends on one single market, the United States (a situation tending to become accentuated), its economic dependency in relation to that country will become greater. Therefore, it is indispensable for Venezuela to do its utmost to attain greater diversification in its markets.

   d. Because of their loss of power and position, it is assumed that great oil companies established in export countries will tend to wish to reinforce their position in industrial nations in the fields of refining, distribution, marketing, and transportation. I therefore feel that Venezuela's government should study and develop—over the short term—programs within the aforementioned fields that would be the most adequate for the country.

It is on account of all the foregoing that a diagnosis of actual problems should be made, and urgent and accurate decisions and measures must not be delayed. Otherwise the consequences could prove to be harsh. Venezuela now has the nucleus of human resources necessary for avoiding adverse situations that could become structural. These resources must be activated harmoniously by Venezuela in order to overcome its present situation.

NINE

The Era of Petroleum Nationalism (1970–1975)

CHANGES IN THE EQUILIBRIUM OF FORCES IN THE OIL WORLD

It is well known that among the different sectors that compose our great human society, there exists a balance of power that at certain historic moments tends to change in space and time, in accordance with the dynamics of several relevant factors: e.g., between governors and governed, employees and workers, public enterprise and private, church and state, parents and children, youth and age, and even within the components proper of each of these sectors.

The decade of the 1960s dramatically revealed a notable change in the balance of power between youth and mature men, the latter in general having exercised up to that time full authority in the leadership of peoples. Due to many circumstances—especially the remarkable decrease in the average age of populations—the youth of most countries was able to increase its influence and power, pacifically in some cases and violently in many others, through a movement known as "the Youth Rebellion," which even today continues to be stirring, and in some cases, quite pronouncedly.

We might also say that an economic and political phenomenon is becoming clearly defined which, like all social phenomena, is a consequence of a process that was incubating previously and which began reaching a crisis at the dawn of the 1970s. I refer to the pronounced decrease in the power of international oil companies in relation to oil-exporting countries. The oil-exporting countries have, furthermore, begun to impose their rights and economic and political viewpoints on developed nations, the great consumers of this vital source of energy.

The history of the petroleum industry in underdeveloped countries of South America, the Persian Gulf, the Far East, and Africa differs in many aspects. However, these countries have numerous characteristics in common. In all of these countries the exploration, production, refining, and sales of this formidable energy resource has in general been in the hands of the same companies, which began obtaining concessions at the beginning of the century. Many of these companies, especially the so-called Big Seven,[1] achieved extraordinary development as production expanded in the various countries, principally Venezuela, those of the Middle East, and, lately, Africa.

Parallel to their economic growth, these powerful companies gained greater and greater control over world markets, whose most important nuclei are the great industrialized nations of Europe, North America, and Japan. It is well known that these companies belong to private investors (or partially to the state, as in the case of British Petroleum) of some of these nations, especially the United States, and on a smaller scale, England and Holland.

As we have seen, since the Second World War markets expanded notably, not only as a result of increased consumption by North America, Europe, and Japan, as well as developing countries, but also because of the displacement of other energy sources, principally coal, by petroleum and its by-products.

The extraordinary increase in consumption of petroleum and the remarkable development of technology and productivity in this field, as well as the increase in prices in most years since World War I and up to the end of the 1950s, caused oil-industry profits distributed among companies and governments to multiply. Up to a certain point, this expansion contributed to the prosperity of several producing countries, such as Venezuela; but above all, it cemented the extraordinary economic power of the oil companies, which, until quite recently, imposed their will on nations and governments.

It is true that in some cases the companies met with extreme reactions by certain governments, such as Lázaro Cárdenas's expropriation in Mexico and Mossadegh's confiscation in Iran, and also with more frequent demands—especially by Venezuela since the 1940s and by the Middle East since the 1960s—seeking a higher state participation in profits through increased taxes applied over the years by parliaments or governments.

---

[1] The Big Seven are Exxon Corp., Standard Oil of Calif., Socony Mobil, Texas Oil Co., Gulf Oil, Royal Dutch-Shell, and British Petroleum Co.

But in general the balance of power tilted toward the oil companies, each time more integrated by means of a network of subsidiaries extending throughout the world, enabling them to exercise great flexibility in all branches of the industry. By controlling the markets, they were able to exert pressure (through boycotts, by decreasing production, etc.) on those countries that resorted to confiscation, expropriation, or any other measure considered by the companies to be contrary to their interests.

Through increased control over production and markets, the oil companies managed to achieve major stability and strength in prices. Prices, up to the mid-1950s, were influenced by the high quotations of the protected United States market (although, as it will be recalled, prices for crudes and by-products of export countries remained well below United States prices).

Toward the end of the 1950s, an excess of supply over demand was caused by the tremendous flow of petroleum from Middle East countries, which hold most of the world reserves and have considerably lower production costs than other countries, such as Venezuela. The entrance of Africa's oil, the pressure on markets exerted by independent producers, and a series of other factors (analyzed in preceding chapters) contributed to this excess of supply over demand. The oil companies therefore applied a policy of reducing closing prices, thereby causing fiscal detriments to producing countries, such as Venezuela, which at that time assessed taxes on the basis of closing prices.

Venezuela, the country that had always been in the vanguard in its relations with the oil companies, showing the way for other producing countries, opposed the price drainage and took the initiative in creating OPEC in 1960, principally to control production and to defend price stability. The oil companies attempted to undermine this organization, which therefore during its first years of existence did not manage to appropriately achieve its aims.

In fact, the Middle East producing countries, due to the very large volume of their reserves, the lack of a sufficiently mature petroleum policy (in turn due to their primitive political and social structures), and increasingly active competition from Africa, were pressured into increasing their production, without being sufficiently concerned about protecting price levels, as tenaciously extolled by Venezuela. On the other hand, the appearance of the so-called independent producers in the market had two principal effects: they contributed to the price drainage and to weakening the international companies' monopoly.

This situation predominated throughout the entire decade of the 1960s. In the previous decade a series of negative international factors had accumulated and reached a crisis, putting an end to a phase of expansion and opening the door to a recession (especially in Venezuela) that was to be felt from the year 1960. Similarly, a number of events occurred in the 1960s that culminated at the beginning of the 1970s in a well-defined and vigorous phenomenon of outstanding importance. Among these events was the creation and development of OPEC; the formation of a higher degree of preoccupation by export countries' governments and peoples over their genuine interests; an increase in energy consumption; and the growth of the independent companies.

The vigorous phenomenon to which I refer is the firm and belligerent policy of nationalism by governments and parliaments of oil-exporting countries, supported in large part by public opinion, in confronting the companies exploiting their minerals and in confronting the great consumer nations. As we shall see, this position has been adopted by all oil-exporting countries (especially in the Middle East, Africa, and Venezuela), with certain pronounced variances and hues but always within—especially through OPEC—a frame of increased demands concerning prices, contractual relations, participation in the oil business, and, finally, the nationalization of the industry. This phenomenon, as a generalized movement, is unprecedented in oil-export history.

It is well known that nationalistic feelings, under one name or another, have existed in all countries and in all peoples since nations were formed and since some became stronger than others. Even before the constitution of states and nations, a similar sentiment was manifested (and we still find it in some scattered tribes of Africa and America) which we might broadly interpret as nationalistic. This sentiment united certain ethnic and social groups in facing other powerful, dominant, and expansionistic groups. Such a spirit gained force and acquired definite characteristics when small European states began to consolidate toward the end of the Middle Ages and beginning of the Renaissance, in a process that extended through the following centuries.

This sentiment, which becomes transformed into outright exasperation when confronting oppression exerted by other peoples, races, or nations, might remain dormant or repressed over a prolonged length of time, even for generations, only to emerge when a series of factors converge and incite its arousal. Such is the origin of nationalism proper.

Throughout history certain leaders were able to channel this spirit, guiding entire populations through titanic battles in search of noble objectives. We have as an example the epic of Simón Bolívar, whose courageous, tenacious, and wise action, aided by other great men and the heroism of many Americans united in a continental ideal of a nationalistic content, achieved the liberation of a major part of South America.

On the other hand, we must also recall that a similarly nationalistic sentiment was a fundamental ingredient causing other leaders contemporary to Simón Bolívar, but of limited vision, narrow objectives, and egotistical ambitions, to destroy the Great Colombia, one of Simón Bolívar's greatest creations. Nor must we forget how this nationalistic sentiment—so natural to peoples oppressed or under stress—has been exploited in all manner and form, even leading along abject and scurrilous paths to the abuse and slaughter of other peoples. A vivid example of this in our present century is the horrendous chapter in the history of Germany, and in the history of humanity, protagonized by Hitler.

It is essential to bear in mind, however, that where Venezuela and other oil-producing and oil-exporting countries are concerned, nationalism does not constitute a phenomenon exclusive to the current decade. A reactionary feeling in facing oil-exploiting companies has existed in Venezuela among many of its citizens since the beginning of the oil industry proper. The fact that exploitation of Venezuela's oil has always been in foreign hands has been the primary reason for Venezuelan nationalistic sentiments. Furthermore, some citizens, such as Dr. Juan Pablo Pérez Alfonzo, upon analyzing several aspects of the industry with a critical and independent spirit, were able to clearly see that oil companies were enjoying excessively advantageous conditions, in detriment to the legitimate interests of export countries.

However, ignorance in the field of hydrocarbons was extraordinarily widespread initially, and furthermore the handful of pioneers—dissatisfied citizens, students of the prevailing situation—had very few possibilities within their reach for influencing public opinion, due to the dictatorial and autocratic regimes then in power. If this was difficult in Venezuela (a nation more advanced than some of the Middle East countries), then we can readily imagine the ignorance prevailing in this field in other export countries and the tremendous barriers that a few knowledgeable citizens had to overcome to be able to freely express their ideas. Profoundly cognizant of the situation, certain citizens have wished to create a constructive line of opposition in confronting it. In order to illustrate this point, I shall briefly refer to the case of Iran.

In Persia (as the country was named up to 1935), nationalism became aroused after a century of exploitation and domination by the British. Abadan was the first oil port of the Middle East, and the most wretched. Up to merely thirty years ago (1944), only one-sixth of Anglo-Iranian Oil Company workers lived in hygienic homes and only 40 percent of all employees were Iranians.

For these and other reasons, outrightly negative to the nation, by 1951 many citizens longed for the nationalization of the oil industry. On March 15, 1951, the parliament voted for nationalization, after a debate which manifested all that had been lost by Iran and all that had been gained by the British, by exploiting the country's petroleum.

Mossadegh, leader of the nationalistic current, became Prime Minister, but he was overthrown in 1953. The Shah then returned to power, but his new Prime Minister had to maintain the principle of nationalization, which he was unable to renounce because of nationalistic feelings that had by then become profoundly ingrained in the people and had even become intensified by results obtained.

After nationalization, Iran received as much income in one year as it had received in the fifty preceding years. In view of this situation, the solution to the oil conflict between the Shah's government and the Anglo-Iranian Company lay in the formation of a consortium, with participation by the so-called Big Seven companies of the world cartel and including those of France and several smaller companies. Likewise, three state companies were created: the National Iranian Oil Company (NIOC), which theoretically was to administer all the nation's oil revenues; the Iranian Oil Exploration and Producing Company, for extraction and transportation; and the Iranian Oil Refining Company, for operating the Abadan refinery.

Thus, in the end, the fair nationalistic movement in Iran protagonized by Mossadegh failed, and he was condemned at that time by the majority of the world press, controlled by the great industrialized nations. Nevertheless, through this unsuccessful movement, Iran managed to increase its participation in its oil business.

Quite frequently some politicians, moved exclusively by ambition and lust for power rather than by truly high aims for their countries, have resorted to demagogy, exploiting the people's nationalistic feelings in facing the manipulation of oil resources by foreign interests. By fomenting xenophobia toward concessionaires for the purpose of obtaining popular favor, they assume attitudes that are only a nationalistic facade, because they merely contribute to exploiting the peoples' emotive feelings without practical or constructive solutions. Such solutions must lead not only to limiting foreign exploitation or removing it, but also to gradually filling the void through programmatic and pragmatic deeds for true national progress.

It is understandably easy to wave the flag of nationalism, but this can have negative consequences if those who hoist it lack any clear objectives of a true nationalistic content, directed through a wise and realistic strategy toward attaining for the nation the management and control of its own resources, without major obstacles or reverses.

Unfortunately some countries have not known, either in the past or in the present, which stand to take and how to clearly answer these fundamental questions: Are they able to undertake alone the great task of exploiting, producing, refining, selling, and transporting their oil resources? If so, how would they carry out this task gradually and yet as rapidly as possible? What phase of the industry, or what proportion thereof, are they capable of handling immediately, using national capital and reliable personnel? Having the necessary funds, how should these funds be utilized? Should national capital be risked (and would it be convenient) in exploration activities in search of new reserves, as well as in other phases of the industry?

In what manner might a program be developed that would enable producer countries to become knowledgeable in—and be able to manage—complicated marketing mechanisms? How, using foreign capital, could certain technical and other agreements be reached, acceptable to the peoples' sentiments, which would not damage nationalistic strategy for the development and expansion of the country and not imply delays or reverses in the industry—which, instead, would guarantee, in addition to increased participation, a secure and rapid evolution toward major and definite control of the oil industry? How might they develop their own oil fleets, with ships built (if technically and economically possible) within their own territory?

Finally, should only the state handle and finance the country's petroleum activities, or would it also be convenient to allow participation by national private entrepreneurs? In the latter case, is national private capital available in sufficient quantity for such purpose, and could it gradually be incorporated in the industry? If such capital is nonexistent, what incentives could be created for promoting the formation of national private savings that might be channeled toward the oil sector? How could private entrepreneurs be encouraged to acquire adequate knowledge of such a complex industry?

Neither these nor many other questions of such vital importance have been systematically formulated within export countries. Nor, much less, have they been answered. Oil countries, their governments, public opinion (or what we might call public sentiment in relation to foreign exploitation of their natural resources), and political and private leaders have been, especially up to the 1970s, more or less passive parties to the powerful and planned action by oil companies and by governments of industrial nations that have traditionally supported those companies. Precisely because of their underdeveloped status, and up to quite recently, oil-producing countries did not have any defined oil policy or clear strategy, or a goal or long-term objective for achieving such ends.

There have been cases where oil countries have taken action, such as the expropriation of the oil industry in Mexico, where a series of uncontrolled and unforeseen circumstances led President Lázaro Cárdenas, in view of the uncompromising attitude of the concessionaires, to satisfy logical nationalistic feelings predominant at the time. However, I have already referred to these events quite fully, and history later proved how lack of planning and of an adequate oil policy prevented Mexico from fully reaping the fruits of this activity. Its formerly great oil-exporting industry has been reduced to a second-rate position and barely covers the needs of its domestic market.

In Iran, we have seen how Prime Minister Mossadegh, by exploiting the justified nationalistic sentiments of the Iranian people and in opposition to the oil companies' inequitable exploitation of the nation's oil resources, managed to nationalize the industry. However, this unplanned achievement by Mossadegh, who did not know how to measure his own forces or, above all, Iran's capacity for handling immediately and alone the administration of its oil industry without adversely affecting it, was the cause, shortly thereafter, of the Iranian government's surrender to the power exercised by the oil companies and by foreign governments.

It is therefore obvious that nationalistic feelings about petroleum have always existed in oil-exporting countries. But collective animosity varying in intensity, according to time and circumstance, has in general been either timidly manifested or repressed; in those sporadic cases where it became vigorously expressed and exploded, as in Mexico or Iran, the expected fruits either did not materialize or collapsed in short time, due to lack of strong, solid foundations, necessary to achieving great objectives.

At the end of the 1960s, conditions were sufficiently ripe to allow the nationalistic tendencies of producer countries to break the barriers retaining them and confront the oil companies and consumer nations. After more than fifty years, the balance of power between the oil companies and consumer nations on the one hand, and oil-exporting countries on the other, was inclined toward the latter. It is therefore possible to set the beginning of the 1970s as the beginning of the era of nationalism in oil countries.

Let us now see how this phenomenon crystallized. In 1969 and 1970, interwoven factors and circumstances that might be called fortuitous, such as the closure of the trans-Arabian pipeline in Syria and the substantial export restrictions imposed by the government of Libya, added to a structural type of factor represented by the already extraordinary and incontainable growth in energy consumption throughout the world, especially in industrial nations, and created an economic phenomenon that was systematically denied by representatives of the oil companies, i.e.: excess of demand over supply. This situation reached a crisis during 1970 and became an emergency in some markets, such as the United States, heavy fuel market, toward the end of that year. Finally, and in the face of such evidence, the companies had to surrender.

On the other hand, the Middle East countries started to realize that an adequate price level would be more convenient to oil-exporting countries than a pronounced increase in production, and furthermore, that joint action by all oil-exporting countries in facing the oil companies could prove to be highly advantageous. Therefore, OPEC finally achieved its true importance as a practical tool to be used by petroleum-producing countries.

By then the oil countries had already acquired a more thorough understanding of a phenomenon that had contributed to aggravating the economic situation of underdeveloped countries, i.e., the frequent decline, in general terms, in prices of raw materials produced by underdeveloped countries, while prices of finished products, manufactured in industrial nations, were being submitted to a process of systematic increase, thereby causing the poorer countries to contribute in a great proportion to financing the increasing inflation in industrialized nations.

I shall later refer to the future steps that, in my opinion, Venezuela should take in relation to its oil business, and more thoroughly analyze what I believe to be a truly nationalistic policy that should be put into motion by the government and nation jointly. I extoll a wholly integrated nationalism, conscientiously structured and planned; neither incipient, superficial, circumstantial, nor undefined in any aspect. The the past, and very especially during the first years of this decade (1970–1975), Venezuelan nationalism has been oriented principally toward achieving a substantial increase in income per barrel for the state and toward striking at the concessionaires through rhetorical attitudes displayed by governors and members of Congress and through a series of laws and decrees, all of which have not constituted any *integrated action* of a nationalistic content. As yet, no replies have been forthcoming to such basic questions as I have posed herein, the answers to which should contribute to the creation of an oil program and strategy characterized by an authentic and rational sense of nationalism.

However, the energetic action adopted by oil-exporting countries since the end of 1970, individually and jointly, has formed the phenomenon that I have been describing as the *era of oil nationalism* . This phenomenon, and the factors contributing to it, have determined a logical counterpart: the weakening, apparently definite, of the oil companies' position in facing export countries.

Although it is not possible to fully predict how events will develop in future years, this shift in power appears to be, I reiterate, of an irreversible nature. This seems to be further demonstrated by the increased dependency of great consumer nations on their oil imports, and consequently on the oil-exporting countries united in OPEC, an organization that is gaining increasing power.

Because of this situation, a consequence of the outbreak of nationalistic feelings in nearly all underdeveloped countries, the oil companies are now witnessing the weakening of their own image. This phenomenon is in fact occurring in many countries, alongside a general deterioration in the image of private enterprise, a phenomenon that might well have negative consequences in developing countries, which require private initiative for propelling their economies.

NATIONALISM IN VENEZUELA—
BACKGROUND

THE EXPLOSION OF NATIONALISM

Let us now see what caused Venezuela, specifically, and the OPEC countries, jointly, to enter fully into the era of nationalism.

We have noted that some interrelated facts occurring in 1969 and 1970 (such as the decrease in production by approximately 800,000 barrels per day imposed by Libya, the interruption in the trans-Arabian pipeline in Syria, and the second Suez crisis), together with the structural phenomenon of the remarkable increase in world consumption of energy, caused a shortage of petroleum. This shortage reached a crisis by mid-1970, instigating a rise in prices of crude and by-products.

In October 1970, the Venezuelan Senate requested Dr. Hugo Pérez La Salvia, the Minister of Mines and Hydrocarbons, to report on the measures being taken by his Ministry for ensuring that new market conditions adequately reflected—for fiscal purposes—the increases in posted prices. For the sake of consistency, I should mention that the 1967 agreement covering reference prices, between the Venezuelan government and the oil companies, ended on December 31, 1971. On November 12, 1970, Dr. Pérez La Salvia reported to the Senate as follows:

> 85 percent of total petroleum volume is transacted through international industry channels and average price levels practically reflect the pricing policy of the oil companies; only 15 percent is applicable to sales to third parties proper. Such a policy has been reflected in Venezuela over the years by a persistently slow decline in closing prices of crude oil: 3 cents per barrel in 1967; 1 cent per barrel in 1968; and 1 cent per barrel in 1969. Likewise, in situations such as that presently prevailing—of a general recuperation in prices to the level of the leading markets and caused by increased demand— Venezuela, as a net exporter, *is not seeing reflected in its sales, in adequate measure, the increased prices being paid by consumers.*[2]

Dr. Pérez La Salvia then stated:

> The difference this year in terms of freights, favoring Venezuela, especially during the second half, is greater. Likewise, our crude and by-products, transported in tankers belonging to groups represented in Venezuela, should

[2]The italics are the author's.

not be negatively affected by exaggerated freight increases, not only because the tankers are owned by these groups but because they make up fleets almost exclusively dedicated to transporting Venezuelan crude and therefore cannot be diverted to the Persian Gulf-European traffic.

Due to the existing pattern of refining, a highly important factor in Venezuelan exports of by-products consists of fuel oil, where a price deterioration can be observed throughout 1969 and during the first six months of 1970.

A certain price recovery occurred during the second half of 1970, but this does not yet reflect the true situation in the world market for this product, due—according to some arguments—to existing long-term contracts as well as to the rise in freights; however, at times of price slumps, Venezuela has always suffered the consequences.

During activities developed in Venezuela's Congress weeks later (December 1970) culminating in a legislative reform, both the government's party, COPEI, and the government proper maintained a passive, expectant attitude; initiative was then taken by the opposition parties. Where the oil companies were concerned, their representatives held overdue consultations concerning the imminent tax reform with representatives of Acción Democrática, COPEI, and high government officials, but obtained hardly any results. Although it is quite true that Congress acted swiftly in relation to the tax reform, it might also be said that the oil companies were unable to adequately foresee such legislative measures or intervene to constructively modify them in terms favorable to the industry.

There are some who believe that the two major concessionaires, Creole and Shell—which hold approximately three-fourths of the nation's total production and therefore occupy foremost positions within the industry (especially Creole)—were not overly interested in avoiding the tax increase to a single 60 percent rate, because that increase, although in turn increasing their own tax load, seriously affected small companies, the "newcomers" as they were described in the oil argot, by eliminating progressive rates and establishing a very high single rate.

It is, however, difficult to know the true position of the companies in this respect, if we consider that the reform not only represented an increase in income tax rates, but also, by empowering the government to establish reference values, placed a powerful weapon in the government's hands for optionally increasing taxes in the future.

In December 1970 an extraordinary international event occurred in the oil world, especially resounding because of circumstances prevailing at the site where it took place. I refer here to OPEC's XXI Ordinary Conference held in Venezuela (Caraballeda). This Conference took place precisely at a time when Venezuela's Congress was discussing the tax reform and the granting of powers to the executive branch for establishing reference values.

It would seem as if the coincidence of this initiative by the Venezuelan Congress and the OPEC meeting was the outcome of a carefully pre-planned and shrewd project. It certainly is true that when the OPEC Conference was convoked to meet in Venezuela in December 1970, nobody even remotely imagined the historical importance that destiny had reserved for it. Much less was it possible to have guessed that for several unforeseen reasons Venezuela's legislative reform would occur precisely when the OPEC Conference was taking place in Venezuela. In any case, it was a fortunate coincidence.

The Conference gained unprecedented importance, becoming a catalyzer of the nationalistic inclinations of its members and a starting point for initiating a joint and vigorous effort channeled toward imposing on international oil companies and on great consumer nations a series of restitutory measures in relation to prices and other aspects.

It was thought—not unreasonably, in my opinion—that in most cases the oil companies did not truly reflect the 1970 increases in the posted prices of crude and by-products in their financial statements for fiscal purposes. This was the spark that caused OPEC's reaction. OPEC, which had been organized for more than ten years, gained strength because its members, in particular the great Eastern Hemisphere export countries, finally realized that by acting vigorously and jointly through OPEC they could achieve the restitutional aims to which they were entitled, thereby tipping the scale of oil power in their favor.

The decisions taken by OPEC in the course of that Conference were:

Resolution XXI-120 covers two clearly defined goals: First, to establish uniform standards for evaluating crudes, and second, to improve fiscal income derived from the oil business directly through establishing a minimum tax rate on oil-company profits and indirectly through increasing posted prices.

Resolution XXI-121, referring to production programs, tends to rationalize member countries' programs and to stabilize long-term prices. It foresees the creation of a permanent committee for the purpose of formulating joint production programs. In this respect, Jahangir Amuzegar, Iran's Minister of Finance, speaking as a representative of OPEC, declared that "all OPEC members are united and shall take simultaneous action." The official communication issued by OPEC stated: "It is obvious that oil companies shall be responsible for the consequences of decisions taken at OPEC's Special Conference, enforcing OPEC member countries' legal rights. This Special Conference has been convoked for Tuesday, January 19, 1971. The Caracas Resolution also contemplates a unanimous decision to be taken two weeks subsequent to a hearing at the General Conference of a Report covering OPEC's conversations with the oil companies."

Resolutions XXI-122, 123, and 125 constitute warnings to industrialized nations and to the oil companies of the principle that OPEC members will enforce their rights when necessary. It is furthermore stated that because prices are highly important to the formation of each nation's fiscal income, they cannot remain tied to United States dollar fluctuations and must be adjusted to compensate for those variations.

Support is also expressed to governments that might take any measures aimed at guaranteeing that the oil companies carry out activities at acceptable levels. It is attempted in this manner to prevent any reprisal by the companies, through reducing their work programs, should they disagree with sovereign decisions by producing countries. According to Resolution 125, such discriminatory measures were applied in the past by the oil companies against member countries when they adopted measures for "protecting their legitimate interests."

It is interesting to notice the oil companies' misguided appraisal of OPEC. Since the founding of OPEC the oil companies contemplated it not only with prejudice and a rolling eye, but—and this was their grave error of appreciation and tactic—even with indifference and contempt. Up to the time of OPEC's energetic action in 1970 (when it was already too late for the oil companies to alter the course of their strategy), they believed that OPEC members would not reach agreement on fundamental aspects. They believed that the Organization would possibly achieve certain results, as in fact it had in preceding years, but that OPEC would never become a powerful instrument for any joint and concerted action by all member countries.

It is undeniable that great differences exist among OPEC members—for example, between Venezuela and the Middle East countries, between the latter and Africa, and even among Arab states proper. Such differences are clearly many, of a historical and even religious nature; even the conditions in the countries' respective oil industries differ. While the Shah of Iran, who governs a country of approximately thirty million inhabitants, burdened with problems, aspired to an extremely high annual production increase, Col. Muammar el-Qaddafi, Prime Minister and strongman of Libya, a country having merely two million inhabitants, enjoyed the luxury of imposing substantial production limitations on concessionaires operating in his territory, ordering a decrease of approximately 800,000 barrels per day. On the other hand, Venezuela—whose proven reserves are by far lower than those of the Persian Gulf nations—could not aspire to any important increase in production.

But the oil companies must at least have suspected that, in spite of enormous differences and mutual distrust existing among OPEC members, sooner or later these countries would realize that they held in their hands the element—petroleum—that is possibly the most essential lifeline for the continued functioning of all consumer nations. The relative scarcity of this product is becoming increasingly critical to the more industrialized nations, all of which depend—to a greater or lesser degree—on the members of OPEC for their oil imports. This dependency is becoming highly pronounced, as is dramatically the case in the United States, because of increased annual energy requirements and the relative decrease in oil reserves. Finally, the industrialized nations will be increasingly compelled to share available resources with underdeveloped countries that are nonproducers of oil.

For many years, and more intensely since the last decade, discontent has been fermenting in underdeveloped countries due to the price weakening of their raw materials. Because of this phenomenon, these poor countries have, in practice, been contributing to the welfare of rich nations. The rich consumer nations, in order to maintain the increasingly higher standard of living of their populations, have continued exporting to the underdeveloped countries, at increasingly higher prices, products manufactured with the raw materials bought from them, in many cases, at depressed prices.

The discontent and unrest brought about by this disagreeable and unfair situation is being felt by increasingly vaster nuclei of populations in underdeveloped countries (stimulated further by their increasing economic problems, which are principally a consequence of their explosive demographic growth, a phenomenon that does not occur in industrialized nations), and this is a fundamental reason for the world's great tensions and crises. These tensions are culminating in a confrontation that is not, as was traditionally believed, between the great blocs of industrialized powers belonging to the capitalistic world of free enterprise, on the one hand, and the communist and socialist world, on the other. Instead, the confrontation is between developed nations, whose populations have an increasing index of consumption and well-being, and underdeveloped countries, whose populations are tending to become further impoverished, partly as a consequence of the unfair structure of international trade that, as affirmed so many times, causes rich nations to become richer and poor nations to become poorer, thus increasing the gap between them.

Because of these circumstances, it was foreseeable that, since the underdeveloped countries possessed the raw material for oiling the industrial machinery on which developed nations depend, sooner or later these underdeveloped countries would make efficient use of these remarkable tools (oil and OPEC). They would finally impose upon industrialized nations, and on the oil companies belonging to them, more equitable conditions, especially concerning prices for their precious product—increasingly scarcer in relative terms.

However, we must not forget that oil-exporting countries have become

Table 9.1   OIL PRICES, 1960–1972

Year	Gulf of Mexico, East Texas (38°–38.9° API)	Venezuela Office (35°–35.9° API)	Kuwait, ex Mena, etc. (31°–31.9° API)	Iran, Kharg Island (34°–34.9° API)
1960	3.25	2.80	1.59	1.78
1961	3.20	2.80	1.59	1.78
1962	3.10	2.80	1.59	1.78
1963	3.10	2.80	1.59	1.78
1964	3.10	2.80	1.59	1.78
1965	3.10	2.80	1.59	1.78
1966	3.11	2.80	1.59	1.79
1967	3.11	2.80	1.59	1.79
1968	3.16	2.80	1.59	1.79
1969	3.32	2.80	1.59	1.79
1970	3.40	2.80	1.68	1.79
1971	3.60	2.80	2.19	2.27
1972	3.60	3.21*	2.37	2.47

*Export values published by the Venezuelan government.
SOURCE: *20th Century Petroleum Statistics,* Degolyer and McNaughton, New York, 1974.

acutely aware of a fact which in all respects is inequitable in the opinion of their leaders, i.e., that regardless of the increasing value of petroleum due to its relative scarcity, United States oil export prices have been high for many years. Through a protectionist artifice, United States oil prices have been—especially during the 1960s and up to 1972—much higher than those at which oil is traded beyond its borders and oil that is, in increasing quantities, imported by the United States. This is evident in Table 9.1.

It is also interesting to observe the much higher participation that great consumer nations have in each dollar of imported oil, in concept of taxes, transportation, refining, and marketing.

In spite of all the foregoing facts, the oil companies underestimated OPEC and consequently disregarded the legitimate interests of oil-exporting countries. As I have explained previously, in Venezuela the oil companies attempted up to the last moment to lead the government, businessmen, and the public to believe that a world situation of excess oil prevailed and that it would continue for many years to come, whereas the truth was entirely different.

Hence, since December 1970, an era of deeply rooted nationalism commenced in Venezuela and in the other oil-exporting countries. Furthermore, this is not a transitory era, but rather a worldwide affirmed reality being projected toward the future in a clear way.

Some people might have possibly believed that these developments were preplanned and that action was taken with a great sense of opportunism. Those of us who have had the privilege of witnessing these events well know that they are not the fruit of foresight, but rather the consequences of a series of circumstances that are linked in a clearly defined pattern. Investigators will also possibly say—in this case quite rightly— that although in the preceding fifty-three years of Venezuelan history, petroleum policy decisions had always been taken by the executive branch, in this case, and in subsequent legislative acts (excepting the gas nationalization law) up to 1973, initiative was taken by Congress.

President Rafael Caldera's Democratic Nationalism

In public addresses during his constitutional term (1969–1974), Venezuela's President, Dr. Rafael Caldera, insistently said that his government was advancing a policy that he described as democratic nationalism, especially in relation to petroleum affairs. In practice, this democratic nationalism was nothing more than a phrased policy of his government, because, as mentioned, the principal measures adopted between 1969 and March 1974 were at the initiative of opposition parties represented in Congress, parties that displayed a markedly nationalistic activity.

Regardless of the so-called democratic nationalism that was widely publicized in official propaganda media, throughout this governmental period Venezuela's oil policy was directed more by the legislative branch than by the national executive branch. It should be noted that the government party lacked a majority in Congress and did not seek any coalitions with the other parties.

Produced through congressional initiative were the law of reversion of assets of hydrocarbon concessions; the December 1970 income tax law reform; and the law reserving for the state the domestic market covering gasoline and other hydrocarbon by-products.

Insofar as petroleum affairs are concerned, Dr. Caldera's government is recognized as having introduced the law whereunder exploitation of the natural gas industry is reserved to the state, but the project submitted by the executive branch was approved by Congress only after incorporation of substantial modifications formulated by the opposition parties.

Possibly where President Caldera's government acted with greatest ability in petroleum affairs was in the handling of reference prices (which were progressively increased) by applying the 1970 income tax law reform, empowering the Venezuelan state to unilaterally establish these prices. Also, at the end of the constitutional period, Dr. Caldera's government acquired two small tankers, *Independence I* and *Independence II,* that may well become the forerunner of the oil fleet that must be formed by the Venezuelan state in the near future.

Venezuela's 1973 Electoral Campaign

In the December 1973 general elections for the Presidency of Venezuela and for governmental bodies (National Congress, legislative assemblies, and municipal councils), petroleum affairs hardly figured in the platforms presented by the candidates for consideration by the electorate. Curiously enough, the candidates made some references to the oil industry only when the campaign was approaching its end. Since Venezuela is such an important petroleum country, with the oil industry so overwhelmingly important to its general economy, it would seem quite normal for presidential candidates to publicly debate their thoughts and proposals concerning the oil industry. However, this again was not the case in Venezuela in its 1973 elections.

Dr. Lorenzo Fernández, the candidate of COPEI, the government's party, possibly was not interested in publicly discussing what had been the oil policy of his outgoing copartisan, President Rafael Caldera. Carlos Andrés Pérez, Acción Democrática's candidate, who won the elections, possibly felt this was a subject requiring complete and in-depth investigation and study, especially relating to its present fundamental aspect—nationalization—which would take place after he gained power.

However, toward the end of the campaign, COPEI's candidate proclaimed that, should he win the elections, the Orinoco Tar Strip, which had been untouched for forty years, would be exploited by Venezuelans; and Carlos Andrés Pérez promised that, should he be elected, the oil industry would become nationalized during his term of government.

THE EVOLUTION OF OIL NATIONALISM UP TO 1974: PRINCIPAL FACTORS

Price Explosion

One of the most important effects of oil nationalitic trends by producing countries, a process which started to materialize in 1970, is the extraordinary rise in oil reference prices, as clearly shown in Table 9.2.

These prices, which oscillated between $1.68 per barrel (Kuwait) and $2.55 per barrel (Libya) in January 1971, have increased more than sixfold, reaching levels between $10.80 per barrel (Indonesia) and $15.77 per barrel (Libya) in January 1974.

Price growth was relatively moderate up to 1973, when several increases occurred and culminated in the pronounced hike of January 1, 1974. In this phenomenon lies the real root of the problem, which the great

Table 9.2   OIL PRICE EVOLUTION, 1971–1974
(Dollars per barrel)

Country	Jan. 1, 1971	Jan. 2, 1972	Jan. 1, 1973	Jan 11, 1973	Jan. 1, 1974
Saudi Arabia (light)	1.80	2.479	2,591	5.119	11.651
Iran (light)	1.79	2.467	2.579	5.341	11.875
Kuwait	1.68	2.373	2.482	4.903	11.545
Abu Dhabi (Murban)	1.88	2.540	2.654	6.045	11.636
Iraq	1.72	2.451	2.562	5.061	11.672
Qatar (Dukhan)	1.93	2.590	2.705	5.834	12.414
Nigeria	2.42	3.176	3.561	8.171	14.690
Libya	2:55	3.386	3.777	9.061	15.768
Indonesia	1.70	2.21	2.26	6.000	10.800
Ecuador	n.a.	n.a.	n.a.	10.000	13.700
Venezuela	2.12	2.99	3.16	7.24	14.080

industrial nations have tried to project as an oil crisis and which has provoked confrontations with export countries.

Oil-price adjustments occurring during 1974 were fiscal adjustments, made in an attempt to compensate for the many years in which this fundamental energy resource had been sold at subsidy prices and, at the same time, in an attempt to reduce profits earned by oil companies. However, the oil companies merely transferred these price increases more than proportionately to consumers, thereby obtaining their highest profits in history.[3] These increased oil-company profits contributed to a further increase in the reference values of crude, established by member countries in the December 13, 1974, OPEC Conference in Vienna. The Conference alerted import countries not to permit the $0.38 per barrel increase in the price of crude—which was to prevail from January 1, 1975—to be transferred by the companies to oil consumers.

The December 1973 Oil Crisis

The energy shortage of the last decade in the United States and in other industrial nations had as a quite fundamental reason the maladjustment between a growing domestic demand and the regular supply of hydrocarbons dependent on foreign sources. It was therefore simply a short-term problem of supply; for the medium and long term it consisted of the

[3]*The Petroleum Situation*, a monthly bulletin of the Chase Manhattan Bank, states that for the first nine months of 1974, performance raised the group's (oil companies') profits for the year to date to $13.2 billion—76 percent more than in the same period of 1973.

insecurity of continuing to rely on those supplies in the face of fear that at any moment they could be interrupted by exporting countries.

However, the December 1973 oil crisis was distinctive due to characteristics that differentiate it from all other crises in the history of petroleum. It will be remembered that in the Achnacarry Agreement, in 1928, it was the great international oil companies' cartel that imposed the petroleum industry's market conditions, but only for the purpose of overcoming the conflict of interests that existed among themselves. In 1973, it was not a matter of settling differences among the big companies, but rather a matter of a confrontation between the cartels of the multinational oil companies and the oil-exporting countries (represented by OPEC), as well as a confrontation between the latter and the group of industrialized oil-consumer nations. Therefore, in this crisis there are several individual parties involved.

Consequently, events such as the embargo on Arab exports, the steep rise in prices, the phenomenon of bulging fiscal incomes (accumulating in oil-exporting countries), and the drainage on the economies of industrial nations are sufficient to demonstrate that the crisis, which started in 1973, is totally different from all its predecessors.

The possibility that exporting countries could resort to interrupting oil supplies to consumer nations—which until 1973 had merely been a reasonable fear—gained force when the Middle East states resorted to measures such as embargoing their exports destined for the United States, Canada, Great Britain, Holland, etc., and reducing crude production in Saudi Arabia, Iran, Iraq, etc. Although the oil embargo was suspended in the following months, there has since been no doubt that oil supplies to industrialized nations could be cut off at any moment—as they were in December 1973—through embargoes of exports by oil-producing countries.

The rise in oil prices since September 1973 is, of course, one of the most important events in this critical period, because although the interruption in supplies subsided and then disappeared—its effect having been limited to a few months—increase in prices has nevertheless a vast future scope. If the energy crisis in the United States was in the past a problem of supply, or of imbalance between supply and demand, it is now, in addition, a problem of inflation, although in smaller proportion.

The increase in prices reverberates in two directions in oil-producing countries: on the one hand it provides them with undreamed-of bulging fiscal income, so vast as to cause profound alterations in their respective economic structures; and, on the other, it affects their inflationary processes, which, on a worldwide scale, have been growing since the 1950s. I shall revert to this factor later.

The situation in less-fortunate Third World countries, those that lack hydrocarbons, is more endangered, because they receive no direct benefit from the increase in oil prices. To the contrary, their economies are weakened, since by paying more for the oil they consume, they are further removed from their possibilities and efforts to become industrialized and developed. In spite of these effects, the opinion still prevails in OPEC that oil prices will continue rising, as announced in its meeting in Vienna of September 12–13, 1974.

The December 1973 oil crisis brought to light the fact that the underdeveloped countries, the so-called Third World, have within their power the possibility of profoundly affecting the economies of developed nations. Beyond any political currents, we can easily perceive the advent of a shift in international trade relations, wherein underdeveloped countries might make use of their privileged position of possessing the raw materials required by the world for its survival; then, industrialized nations will become dependents, as they already are becoming, where petroleum is concerned.

Today's international equilibrium might also yield to another structure of forces. All is possible, because the negative consequences of inflation can be corrected and controlled through normalization and subjection to special conditions of reciprocal trade between developed and underdeveloped countries. We must not forget that countries exporting oil—and raw materials in general—are the foreign markets for the products manufactured in industrialized nations, who must protect those markets from penetration by their competitors.

The Situation at Present

Since the so-called oil crisis that started in September 1973 still prevails in some fundamental aspects that will probably continue for a long time, the current situation in the scope of this book is represented by events occurring in the world since the beginning of 1974, which I shall now summarize, as I consider them to be of singular historical importance and also of significance to the future of Third World countries.

When the increase in oil prices occurred, great consumer nations reacted fundamentally by rationing consumption; by discrediting export countries through publicity campaigns; and by engaging in activities aimed at forming a bloc integrated by consumer nations.

The governments of industrial nations affected by the increase in oil prices adopted transitory measures regulating, toward the end of 1973 and in the first months of 1974, consumption of products derived from petroleum, thereby limiting the normal development of certain activities of daily life. In several countries, circulation of vehicles was restricted, speeds were limited, public parks were closed, electricity was reduced in public buildings, neon signs were dimmed, and supplies of fuel oil and gasoline were reduced.

In the industrialized consumer nations, a sustained campaign discrediting oil-exporting countries (especially the Arab states) was launched, leading the public to believe that they were the culprits responsible for the inflationary process in the world economy and that the rise in prices of crude and its by-products was the cause of the world economic recession.

I believe this opinion is erroneous and also detrimental, because it projects a distorted image of oil-exporting countries. It has been the industrial nations that have over the years been transferring their local inflation to the Third World countries, as a result of international trade based on unequal terms. Also, oil prices started increasing only in recent years, to begin to compensate for the progressively higher costs of products manufactured in industrial nations in recent decades.

The price of a barrel of Venezuelan crude in 1950 was $2.09 and in 1970 only $1.85. This means that over a period of twenty years its price decreased by $0.24, whereas prices of many industrial products increased by up to 100 percent during the same period. Therefore, it seems that neither petroleum, nor petroleum-producing countries, are the principal factors determining the present world inflation, and much less are they the factors causing the economic difficulties that industrial nations are experiencing.

There are many who define the world situation as alarming because a few underdeveloped countries are able to affect the economic structure of developed nations and are on the way to becoming economically independent. There are others who think that the industrial nations, accustomed as they are to imposing their conditions on the rest of the world because of their greater economic, technological, and military power, might resort to force with the intent of subduing politically and militarily these underdeveloped countries.

There are also those who believe that by forcefully attacking these developing countries, the industrial nation would be exercising a defense of their traditional interests. However, should an armed intervention occur, the consequences would be totally negative. They could lead to a confrontation between the world's Eastern and Western powers, or to an understanding between them for imposing their terms on Third World countries, from which nothing beneficial would result for the immense majority of humanity and which could, furthermore, lead to catastrophic events difficult to predict.

Should the great nations now attempt to overcome their economic difficulties by resorting to armed power, as they have done in the past, not only would they be acting outside of the basic principles of coexistence, cooperation, and trade called for in today's world politics, but they would be forcefully overriding the command of reason and delaying, for who knows how long, the building of a more balanced and just international order to which all humanity aspires and has been awaiting for many centuries—an order in which all nations on earth might find their rightful place. Both imperialistic policies and imperialistic economies should give way to a structured system of balanced international trade.

The International Energy Agency

Following the policy of forming a bloc of consumer countries, a resounding event was the creation of the International Energy Agency (IEA) in Paris in 1974 by sixteen nations: the United States, Japan, West Germany, Great Britain, Italy, Canada, Belgium, the Netherlands, Spain, Sweden, Austria, Switzerland, Turkey, Denmark, Luxembourg, and Ireland. (France and Norway declined to join, owing to policy disagreements.)

The main objectives of the IEA are:

1. To assist any member of the group that might suffer a decrease in supply of at least 7 percent, due to any action attributable to producing countries. For covering such an emergency, Agency members agree to reduce their own consumption.

2. To seek and develop new sources of petroleum in countries that are not OPEC members.

3. To study methods that will permit a reduction in petroleum consumption.

4. To develop other sources of energy.

From the aforementioned points, it is quite clear that the objectives are channeled toward neutralizing or diminishing OPEC's power, through mutual cooperation among consumer nations. It is not possible to predict the future of this new organization, because its survival is subject to the degree of cooperation maintained among its members when, under any circumstance, oil supplies to any of its members are decreased. Initially, for example, it appears that Japan is not inclined to share its oil supplies with the United States in the event that exports of crude to the United States are decreased.

Another objective of this new organization is to reduce consumption in member territories, which is difficult because they are industrialized nations and by their nature as such are subject to a process of expansion that will require an ever-increasing demand for oil. Denis Healey, Chancellor of the Exchequer and Governor of the IMF for the United Kingdom, at the Joint Annual Discussion held on October 1, 1974, stated: "But however successful we may be in reducing our demand for oil by conservation, if we were to limit our imports of oil to what we can pay for year to year by exports, we would create bankruptcy and unemployment on an unprecedented scale."

On the other hand, the creation of the International Energy Agency, as an organism to counterbalance OPEC, would naturally provoke reactions in the latter tending to neutralize activities developed by this representative body of consumer nations. According to my preceding considerations, and although this is not a prediction, I might say that the future of the International Energy Agency depends fundamentally on the petroleum policy adopted by all oil-exporting countries, members or not of OPEC. I furthermore feel that members of both organizations should become aware of this fact.

Other Plans

In addition to the creation of the International Energy Agency, other recommendations have been proposed, including measures concerning the recycling of petrodollars and financing of oil purchases.

Among them I might mention the Witteveen Plan (named for Johannes Witteveen, Managing Director of the International Monetary Fund), which recommends the creation of an oil facility at the IMF, to lend funds to oil-consuming nations; the Kissinger Plan, which proposes a $25 billion lending agency to the consuming nations; the Van Lennep Plan, which is similar to the Kissinger Plan, except that loans would be guaranteed by the OECD; the Wilson-Roosa Plan, which proposes a mutual fund for OPEC countries with guarantees against losing holdings through nationalization; the Enders Plan, which recommends higher domestic oil prices in the United States, to continue pressuring for development of alternative energy sources; the D'Estaing Plan, which proposes a tripartite conference between oil producers, oil consumers, and developing countries; and the Shah's Index, which proposes linking oil prices to twenty-five or thirty commodities, in order to encourage oil users to check their inflationary trends.

The United States Presidential Policy concerning Oil

When Richard Nixon resigned and Gerald Ford became President of the United States in August 1974, developed nations had already felt the strong impact of the increase in prices for petroleum and its by-products. Likewise, the economic advantage that the new situation represented to the multinational oil companies had already become evident. Consequently, the national and international petroleum situation was one of the main subjects occupying the immediate attention of the United States' new President.

Important facets of President Ford's oil policy were covered in his speech at the United Nations General Assembly on September 19, 1974, and also in other subsequent public appearances, such as the IX World Conference on Energy Affairs, held in Detroit on September 23, 1974. In his public appearances since his speech at the United Nations General Assembly and up to December 1974, the most outstanding points of Gerald Ford's oil and energy policy orientation in general were the following:

1. To limit domestic oil and energy consumption in general, and to guarantee a reduction in oil imports through prohibitively high tariffs or outright quotas.

2. Proposals for boosting supplies: offshore oil development, exploitation of oil in Alaska, amendment of clean-air laws, and development of the necessary technology for creating alternative sources of energy that will enable the United States to become self-sufficient and thus avoid dependency on other countries for such supplies.

3. Joint action by great consumer nations for counterbalancing—as far as possible—action by OPEC countries and for providing mutual assistance in the event of new drastic decreases in oil supplies.

Some of these goals, such as the joint action by consumer nations, may find support in United States public opinion, as they tend to favor individual interests, as well as the interests of other industrial nations. Basically, however, what is being pursued now is a drastic decrease in present oil prices. The announced objectives are an endeavor to project an image of a coherent policy for achieving this fundamental aim. Actually, a decrease in consumption over the short term could be feasible only in a small unimportant proportion, because a decrease on a larger scale would seriously affect the industrial life of these nations.

In reference to the development of alternative sources of energy, it must be observed that although this goal is very plausible, most of the programs in this field—notwithstanding the continuous propaganda concerning the so-called independence program that commenced toward the end of the Nixon administration—at present have hardly any impetus.

Another fundamental aspect in President Ford's declarations lies in concepts reflecting an ideology contrary to the interests of Third World countries, among which are the oil exporters, including Venezuela. In effect, President Ford accuses oil-exporting countries of being the cause of an inflation that is affecting the economy of consumer nations and, likewise, of the difficulties arising from the increase in oil prices that underdeveloped countries will have to face. In other words, the intention thereby is to establish the United States as the defender of both consumer nations and underdeveloped countries and then to use both groups against OPEC.

Repercussions in Oil-Exporting Countries

The first reaction in oil-exporting countries to President Ford's speech at the United Nations General Assembly on September 19, 1974, was the public letter addressed by the President of Venezuela, Carlos Andrés Pérez (who took office in March 1974, for a five-year period), to the United States President on September 20, 1974, This letter was transmitted throughout the world.

The document, although classified by its author as the reaction of the Venezuelan government to plans by the United States government, expresses the feelings and opinions of other underdeveloped countries as well. Therein, Venezuela's President rejects President Ford's accusations against oil-exporting countries as being attempts against their sovereignty; he shows the lack of foundation for such assertions and clearly demonstrates that it is precisely the developed nations that are the originators of the world's prevailing unfair economic structure, as well as of the underdevelopment and hunger endured by Third World countries.

In his observations to the President of the United States, President Carlos Andrés Pérez points to the share of responsibility corresponding to industrial nations in relation to the economic situations prevailing in underdeveloped countries. In summary, and for the appraisal of my readers, I shall reproduce hereunder some paragraphs extracted from the letter:

1. We in Latin America, as in other developing countries, are able to state that industrial countries have been abusing the fundamental needs of the Latin American, Asiatic, and African peoples.

   To mention the particular case of Venezuela, oil prices were, during many years, openly deteriorating, whereas we were obliged on the other hand to receive manufactured products proceeding from the United States at increasingly higher prices which, day by day, further limited our possibilities for development and well-being.

2. The food crisis throughout the world is, among other reasons, the consequence of the high prices at which developed countries sell to us their agricultural and industrial machinery, and other needs essential for our agriculture and the growth of our economies.

3. The creation of the Organization of Petroleum Exporting Countries (OPEC) was precisely the direct consequence of the employment, as an instrument of economic oppression by developed nations, of a pricing policy unworthy of our raw materials.

4. Ours have always been the countries weighed down by unacceptable burdens in international commerce.

5. Our complaints and claims went unheeded and our legitimate aspirations were ridiculed.

6. . . . oil prices barely affect in a minimum percentage production costs in the United States and in other developed countries.

7. Venezuela will certainly applaud all intents to solve the great problems of our time, in global terms, providing such a world perspective does not signify a predominance of great countries over small countries.

8. I am taking the liberty of interpreting OPEC's policy in stating that oil producing countries aspire to achieve, within a world frame such as that of the United Nations, an equitable level of understanding and of international justice between raw material producing countries and industrialized countries.

Such points of view, contained in President Pérez's message, are of interest not only to Venezuela because they express the country's views, but also to all countries making up our international community, be they rich or poor.

In this communication, President Pérez also expresses political and economic ideas that are the concern of our time at a world level, and he includes a formula—a trade alternative—for facing and solving, in practical terms, the international imbalance in which traditionally most of the earth's people have lived, divided today into rich countries and poor countries.

No other subject has more occupied and preoccupied the minds of men—leaders and/or led—whatever their country may be. The problem of the so-called economic and technological dependency, being today debated at international summit meetings, and up to a short time ago a matter of individual concern to underdeveloped countries, is now of interest to, and affects in a high degree, all developed nations. In other words, this subject has now become a worldwide controversy. It is no longer an isolated debate relating to the private conveniences of developed or underdeveloped countries and to the degree of injustice in which trade operates between them. Rather it has become a search for a solution that will effectively balance international relations in the world of today and of the future.

It might therefore be said that the message communicated by the President of Venezuela has an immediate and opportune meaning as a reply to the United States President's speech; and furthermore, it has a permanent value because of its varied and valuable implications.

For Venezuela, Carlos Andrés Pérez's message contains the viewpoint of what is, and what should be, the basis of its present oil policy, especially in relation to price factors and to Venezuela's support of all OPEC members.

For Latin America, the message contains a true and sincere focus on the state of trade relations between the United States and this part of the world, summarized in its negative effects: "the impoverishment of our countries as tributaries tied to the United States economy." As a determining cause of such regrettable relations, Presidnet Pérez points to, among others, the lack of comprehension and understanding between the United States and countries in development, in facing the need to seek adequate formulas for equal treatment and mutual economic respect.

Historians, experts, and students of the political and economic causes that lead to the existence of developed nations in the world exercising a subjugating predominance over others, have maintained that in order to comprehend the problem, it is essential to express it with a criterion, and in the language proper, of the underdeveloped countries, reflecting the economic ties that have bound them and to lessen the distored image projected by their people in developed nations.

In his message, Venezuela's President dramatically and objectively expresses the viewpoints and motivations of underdeveloped countries, producers of oil and raw materials in general, and in their name he categorically denies the charges made against them by industrial nations: "The economic struggle has originated through the large developed countries' denying equal rights to emerging countries seeking essential equilibrium in international trading terms."

Finally, the message contains a challenge to all Venezuelans in particular, and to all countries that today belong to the so-called Third World, for them to engage in the noblest of all modern struggles for independence: to suppress economic colonialism and to unite the bonds of dependency that enchain them to the small group of developed nations.

The future of Venezuela's petroleum industry is naturally closely tied to that of other exporting countries. Let us now examine how Venezuela can, and must, construct the future of its oil industry, and hence, the future of the whole nation.

PART THREE

The Future

TEN

Venezuela:
A Truly Petroleum Nation

INTRODUCTION

To the present time, Venezuela—like all petroleum-exporting countries in general—has not been a truly petroleum nation, but rather a country having a wealth of hydrocarbon resources exploited by foreign companies. With the minor exception of CVP (the Venezuelan Petroleum Corporation, established in 1960), Venezuela has in general been satisfied to passively receive its oil revenue and systematically claim an increasingly higher participation per barrel. In other words, it has merely acted as a state receiving funds proceeding from such an industry, exploited and administrated by foreign companies.

The following remark by petroleum expert James Akins (Director of Energy Affairs for the U.S. Department of State during the Nixon administration) is significant: "There have existed—and still do—countries by far richer than any OPEC country, but none so small, so intrinsically weak that have earned so much in exchange for so little self effort. . . ."[1]

The era that Venezuela is now entering is precisely one of a truly

[1] James Akins, "The Oil Crisis: This Time the Wolf Is Here," *Foreign Affairs*, April 1973, p. 481.

petroleum nation; a nation which, having the privilege of possessing substantial volumes of a source of energy increasingly more essential to humanity, should finally and clearly see its own possibilities and start using its own initiative in order to assume an imaginative, enterprising, audacious, and truly nationalistic position, based on a thoroughly defined and modern program of activities, using a shrewd strategy aimed at injecting its oil industry with a good dose of dynamism for achieving the expansion and boom proper to a truly petroleum nation. It shall thus be able to increasingly take over the responsibilities of handling and controlling its oil industry, studying and executing formulas and associations that will lead to a high degree of expansion. However, to attain this expansion Venezuela must bear in mind that time is limited for taking advantage of this opportunity to become fully developed in all economic, political, social, and cultural aspects.

If Venezuela uses the extraordinary oil funds it is currently receiving to achieve its high national goals, i.e., the creation of a vigorous economy and formation of a healthy and educated populace having a high income per capita distributed uniformly, then it will, in the fairly near future, attain sufficient diversification, economic stability, and a social and cultural level that will permit its self-sustained growth, even when its petroleum production irreversibly declines after its deposits have been fully exhausted.

We have been able to see throughout this book that the oil business is eminently international. However, its internationalism has always been monopolized by a handful of companies that—through their certainly praiseworthy activities, in some cases even worthy of imitation—knew how to find petroleum in many regions and how to produce, refine, and transport it satisfactorily, but to their own advantage and to the advantage of their industrialized homelands. Therefore, I feel that Venezuela should now endeavor to secure its position as a truly petroleum nation by directly and actively participating in production, refining, transportation, and marketing phases and even extending these activities into other countries, through CVP and other wholly owned national and international companies.

Consider a nonpetroleum country, such as Italy. Through the remarkable capability of mainly one man, Enrico Mattei, Italy was able to develop its own international company (ENI), which even achieved predominance in world oil affairs. Thus it seems to me incomprehensible that Venezuela, being one of the most important world petroleum producers still possessing rich deposits and specialized personnel in the field, has not yet created and developed the appropriate and necessary organization for directly operating the various phases of its oil industry at both national and international levels.

BASES FOR FORMULATING A COHERENT, AUDACIOUS, AND GENUINE VENEZUELAN PETROLEUM POLICY

We must plan with audacity, and execute with prudence.   SIMÓN BOLÍVAR

Letter to General Francisco de Paula Santander, Cúcuta, June 25, 1820.

The unprecedented prosperity favoring all petroleum-exporting countries, the eager markets, and increased oil prices are propitious factors for Venezuela to immediately concentrate its efforts on becoming a *truly petroleum nation.* Venezuela still possesses considerable petroleum resources and should therefore develop and execute a vast program aimed at taking over the administration of its oil industry in a practical, rapid, and systematic manner with a view to achieving international participation in world petroleum affairs.

As a first step, all phases of its hydrocarbon industry should be nationalized; second, its petroleum resources must be expanded; and third, it should strengthen CVP's structure, as well as that of all other companies that might develop from nationalization, in the technical, administrative, and financial fields.

Prior to this English version of my book, on many occasions I supported the advisability of nationalizing Venezuela's oil industry, even before September 1973, when favourable changes arising from increased petroleum prices started to take place. However, there were those who believed that nationalization would not be feasible because Venezuela was not yet technically and administratively equipped for directly exploiting its petroleum industry without a certain degree of outside assistance.

This argument is debatable, since the oil industry is an international activity and obtainment of external collaboration should always be feasible. Furthermore, Venezuela might even, along the way, form its own technicians at specialized training centers at home and abroad.

Another opposing argument was that the country did not have the necessary surplus funds for expanding mining-risk investments for explorations for new petroleum deposits. In recent years national budget expenditures were carrying a deficit, and there are great social and economic needs prevailing in the populace.

Although in those days this was a very realistic reason, it was nonetheless rejected by nationalization partisans, such as myself, on the basis that the petroleum industry could at least partly generate the funds necessary for meeting its own financial requirements and, in all events, could rely on sufficient external financing. In this respect I stated that, contrary to what has frequently been maintained (interpretations having been misled due to the shallowness of local analyses related to petroleum), private capital existed in Venezuela and was being generated in sufficient volume for participating in certain petroleum projects. This was fully evident by statistics showing the vast sums of money which each year—even before 1973—and in increasing quantities, were monopolizing bank deposits and mortgage bond investments and circulating in other public and private liabilities. It was evidenced by the millions of dollars remitted abroad to secret bank accounts in Switzerland or to United States banks to be invested at fixed interest rates, or placed in mutual funds or shares in the United States financial market, not to mention the excessive and alarming volume of liquid funds frequently available in Venezuela.

The preceding facts therefore clearly prove that voluminous private capital did exist (and still does) which could be oriented toward the petroleum industry once it became nationalized. In any event, at present (1975) Venezuela, besides having sufficient resources for entering programs aimed at expanding its oil industry, has already started investing its excess funds abroad.

The preceding arguments supporting nationalization of Venezuela's petroleum industry gradually penetrated local public opinion, to such an extent that the process of nationalization would have taken place regardless of the extraordinary income of oil money generated through increased petroleum prices. I certainly do not deny that this circumstance greatly influenced optimism favoring nationalization, but the tendency already prevailing in the petroleum policy and in public opinion would have nonetheless led to nationalization of the industry. Most Venezuelans today feel that the climate is now propitious for nationalizing their petroleum industry; all business sectors unanimously approve the measure; concessionary companies are serenely accepting it as an irreversible fact; and the government has finally decided to carry it out at the soonest feasible moment.

I shall now refer separately to the nationalization of Venezuela's oil industry and to the strengthening of CVP, which are the fundamental bases of Venezuela's petroleum policy.

Nationalization

Nationalization of a powerful petroleum industry in any country is, of course, a vast and intricate process that cannot possibly originate from a mere whim. It must be strictly based on thorough, careful, precise, and detailed programming. Venezuela's government is channeling nationalization through its legislative branch, subsequent to a prior study of the process and to the formulation of proposals concerning the manner in which it should be executed. In his speech of May 16, 1974, President Carlos Andrés Peréz announced the designation of a Committee for Reversion, with representatives from all national sectors. This Committee will be responsible for studying possible operative alternatives of maximum advantage to the country, once the petroleum industry has become nationalized. The President likewise described the following basic conditions covering nationalization procedures:

1. Compliance with Constitutional mandates, paying fair indemnities to concessionary companies, but not exceeding prevailing assessed real values—less amortization—of assets that shall, in effect, revert to the country.

2. Nonacceptance, under any concept, of asset revaluations, nor of any other form of accounting or administrative maneuvers.

3. Deduction of whatever sums are required for fully guaranteeing payment of all fringe benefits and other rights to which workers and employees are entitled.

4. Deduction of all amounts owed to the National Tax Bureau, for any concept.

5. Payment in Public Debt Bonds of agreed indemnities resulting from prior studies approved by Congress, under appropriate terms most advantageous to the country's interests.

As nationalization shall be a comprehensive procedure, barring separations or discriminations, the process includes all Service Contracts which may be resolved either through agreements between CVP and the contractors, or through expropriation.

Other relevant aspects concern decisions that will secure, in practice, the uninterrupted continuity of all activities within the petroleum industry.

The State shall become the proprietor of company assets, but not of the companies proper. Consequently, new companies must be created for the reception of assets involved and for executing all administrative and managerial activities.

The President furthermore added that:

The depth and intensity of the necessary preliminary studies are quite obvious; however, in principle it would be unwise to transfer all individual unit activities, at present operated by these companies, to a single national company, as such a measure might possibly bring catastrophic consequences to the normal continuity of such activities, which at all costs must be fully guaranteed.

These new companies must be headed by Venezuelans who have had experience in the administration and management of the international companies, thus assuring the smooth operation of the State's entrepreneurial efforts.

In connection with the preceding official statements, I personally wish to add that the process of Venezuela's petroleum nationalization should pursue the following general objectives:

1. To secure and increase income proceeding from this vital industry for the nation.

2. To secure an administrative and operational system, similar, or superior, to that existing at present. I stress the necessity to rapidly form a legion of Venezuelan professional and administrative petroleum experts, capable of conducting the national and international aspects of the business in all its phases.

3. To elaborate a program of investigations leading to the formation of technological self-sufficiency in all phases of the petroleum industry and in those of its collaterals, such as gas and petrochemicals.

4. To assure that pure and applied research shall provide technological self-sufficiency to the nation, in the shortest possible time.

Finally, another aspect of great importance is that any country desiring to nationalize its petroleum industry must ensure that the process does not end by bringing a windfall to the nationalized international companies, as has happened in some cases. We must not forget that the nationalistically tinted measure adopted in 1973 by petroleum-exporting countries to increase petroleum prices, although bringing higher revenue, in actuality favored the multinational oil companies, which were able to substantially increase their profits.

Venezuela's government must be especially careful in preventing such a possibility when nationalizing its petroleum industry, because however strange it may seem, when concessionaires are cut off from direct exploitation of the industry, they might slide into a more advantageous position in world petroleum affairs. This alternative favoring the oil companies might in practice occur, if on the one hand they are relieved from a series of burdens, operational expenses, obligations, and controls and on the other hand they are able to secure supplies and marketing of petroleum, thereby gaining further international power insofar as transportation, refining, and trading are concerned, in addition to their technological skill. Hence, all these possibilities must be taken into account, because should they occur, they will prove to be detrimental to the scope of Venezuela's nationalization process.

Strengthening CVP

The other basic objective of Venezuela's petroleum policy consists, as mentioned, in fortifying CVP and other companies that may be formed or transferred to the state. This aim has always existed, but it is now essential and nondeferrable. These state companies shall be responsible—with due assistance and technological collaboration when necessary—for fully developing the areas assigned to them. They shall be responsible for the execution of a vast exploratory program, for creating their own adequate refining capacity, for efficiently executing domestic and international marketing affairs (offering Venezuelan consumers a service at least equal to that provided formerly by private companies), and for efficiently taking over the management of concession areas and installations that will revert to the nation through nationalization.

To be able to effectively perform these responsibilities, state companies will have to enter into service contracts, create companies in association with private Venezuelan capital, and give incentive to the formation of private Venezuelan companies for providing miscellaneous services to the industry.

On an international plane, alongside the wide range of activities carried out directly by state-owned companies, especially where marketing is concerned, a network of subsidiary companies should be created (owned by these state companies)—such as VENFUEL, already created by CVP and constituted in Coral Gables, Florida, and FUELCO, constituted in the Bahama Islands, as well as other joint-capital companies. Finally, the formation of Venezuelan private companies to operate internationally should also be stimulated.

The ultimate goal of the joint action developed by this group of companies would be to gradually encompass the vast international petroleum business, starting from straightforward and simple capital investments in foreign companies and continuing up to eventual exploitation of oil deposits in other countries.

The bases I have referred to for formulating a coherent and audacious petroleum policy seem to exist in the orientation of Venezuela's present government, at least where the decision for imminent nationalization of Venezuela's oil industry is concerned.

Appropriate Orientation for Participation by Foreign Capital and Technology: Service Contracts and Mixed Companies

The so-called propitious climate for investments in most Third World countries has progressively deteriorated. Many of them, expecially in Latin America, are closing their doors to direct foreign investment and restricting participation in certain industrial and service sectors. Countries still fully receptive to multinational companies and to direct foreign investment in general, such as Brazil, are today an exception rather than the rule.

In outlining the possibilities for appropriately orienting direct foreign investment, especially proceeding from multinational companies, we must bear in mind the basic business nature of these companies, which obliges them to offer attractive dividends to their shareholders. They would otherwise tend to disappear, being unable to nourish their respective social capital or to obtain financing under favorable terms. It is thus that these organizations, as well as private enterprise in general, besides the contributions they afford to the community and country in which they operate (currently higher than expected and demanded of them in the past), have the fundamental mission of coordinating human, capital, technological, and administrative resources.

Taking into account the aforementioned negative attitude by many receiving countries where direct foreign investment is concerned, the only alternatives would be to definitely eliminate activities by multinational corporations and foreign investment in general or to gradually restrict them[2]—which seems to be the aim of many governments in underdeveloped countries or to find new cooperative formulas advantageous to all parties. Where this last alternative is concerned, I feel it will be necessary to seek constructive solutions related to one fundamental goal: ownership control. In fact, increased preoccupation by emerging countries over protecting the the integrity of their national sovereignty and autonomy is closely tied to this concept, because ownership control by foreign companies is deemed injurious to that integrity, which is today of more importance than ever to these countries.

Nationalistic sensitivity resents control by foreign interests, and this is the origin of the increased discontent generated by direct investment. When such conflicts occur, and whatever the reasons supporting the parties involved in each case may be, the receiving countries feel unfairly treated, and it becomes difficult to find appropriate solutions. Consequently, reactions frequently tend toward a confrontation that eventually leads to state intervention, expropriation, or even confiscation and other measures that could point to a lack of knowledge of the rights acquired by foreign investors.

Since the beginning of the 1970s, we have witnessed a change in the equilibrium of forces between petroleum companies and exporting countries. The latter's position has increasingly strengthened, and these countries are now vigorously imposing their terms. This change, favorable to the host countries, is now becoming generalized and felt in other areas of foreign investment, e.g., in commercial banks established in Latin America and controlled by foreign interests.

I might mention here three sectors that have traditionally supported or accepted foreign investments: the Armed Forces, the Catholic Church, and private national entrepreneurs. Lately, in many underdeveloped

[2]Cartagena Agreement Resolution XXIV, regulating foreign investments in the Andean Region.

countries this attitude has now changed. In fact, a nucleus in the Church has shifted to the left, and in several countries this has also occurred in the Armed Forces. Private national sectors today tend to view the great multinational corporations with apprehension and are inclined to seek protection against competition by these companies.

This tendency has been accurately described by Peter Gabriel, Dean of the Faculty of Economics of Boston University:

> Judging from the convulsions today shaking most underdeveloped countries, this tendency (nationalistic xenophobia) is inevitably gaining force due to economic and internal policy reasons. In effect, if we consider the increasingly precarious position of foreign companies operating in the Third World—either in the agitated Andean Group of countries; in Asia; in emergent African countries; or in the rich oil lands of the Middle East—it can only be concluded that the era of multinational corporations, as traditional direct investors in less developed countries, is reaching its end. Consequently, possibly a tendency leading to the entire termination of their role in economic developments might occur, unless they are able to adapt themselves to operating under institutional arrangements different to those that were so advantageous to them in the past. From the receiving country's viewpoint, any acceptable alternative method allowing importation of foreign investment must be devoid of the disadvantages involved in conventional direct investments.
>
> Nearly all such disadvantages are related to ownership rights which, theoretically, extend into perpetuity. These rights are the conceptual essence of direct investment. They preoccupy receiving countries because they perpetuate foreign control over local industries, over expanding social capitals, and over the increase in foreign currency liabilities.
>
> This is the root of the problem; no country objects to the entrance of private capital as such, nor does it object to the affluence of administrative techniques, of new technologies and of other know-how introduced by foreign corporations. The point in question concerns the bases on which these resources enter. The problem, from the viewpoint of the receptor country, is how to separate the flow of resources accompanying direct investment, from the damages arising from foreign ownership rights inherent to such investments; one might generalize by asserting that even when foreign investment is no longer accepted under the form of a multinational corporation as direct investor, such corporations shall still be well received as suppliers of services.[3]

We have here, then, a possible solution of special interest to the petroleum business, i.e., foreign control over the industry, originating from ownership rights, could be replaced by services, and even capital, supplied

[3]Peter P. Gabriel, "MNC in the Third World: Is the Conflict Unavoidable?" *Harvard Business Review,* July–August 1972, pp. 97 and 98. Gabriel was until recently the principal partner in the well-known United States advisory firm, McKinsey & Co., Inc.

by multinational corporations, on other bases not implying proprietary control. This assumes a change in the object of investment by foreign corporations; it signifies the substitution of an intangible asset for capital, as the primary and essential element. This intangible asset, of great importance, is administrative, managerial, and technological know-how in all its aspects, and without limitations. Consequently, direct foreign investment based on ownership rights would be replaced by contractual agreements with foreign companies, implying their obligation to contribute in the administrative and technological fields by supplying services that could be accompanied by capital but would not represent thereby any control over the company established in the receptor country.

The preceding concept is not at all new. Numerous agreements—in practice functioning satisfactorily—already exist in the administrative, productive, and technological fields. Today we talk of a deficit or surplus in the balance of technological payments in all countries, which in many cases applies to the supply of services, without their necessarily being tied to the direct investment involved. The novelty, therefore, would lie not in the structure as such, but rather in the massive use of the concept proper, with a view to overcoming the grave problems originating through foreign control over property and thus eliminating the risk that the receiving nations' governments ignore ownership rights acquired by foreign investors. Such risks would be prevented, or very much reduced, because the marked nationalistic attitude of Third World countries is, among other things, a reaction to property control, and hence against monopoly by foreign investors, with all the economic and even political consequences of dependency involved.

The Soviet Union is increasingly using this type of cooperative formula, as we have seen by the Fiat contract for building the Togliatti automobile plant, which according to estimates represents $50 million in fees to the Italian company. In Venezuela the development of service contracts, begun timidly a short time ago (1970) after many years of study, involves the multiple and varied possibilities of this system. I therefore feel that appropriate formulas should be devised in order to ensure that the Venezuelan Congress, or whoever else is concerned, does not detain or delay the carrying out of any such contracts that may be signed.

There is another greatly important instrument capable of being successfully substituted for the traditional formula of direct investment (which, in some cases, may be closely related to service contracts) and having a wide margin of application in the petroleum business, which is so much in need of foreign technology and capital. I refer to mixed companies, where the investment of national capital (state or private, or both) is higher than foreign investment. This structure, which allows multiple combinations, would orient the necessary technological and capital resources toward the petroleum sector, with national control being maintained over the respective companies.

Once Venezuela's leaders have acquired a clear concept of what a truly petroleum nation is, they should then aim their activities not only toward producing, processing, selling, and transporting petroleum, but also toward participation in these same activities in other zones of Latin America, and even in the Middle East and Africa. To that effect it would be necessary for CVP, and other future companies, to enter into contracts guaranteeing increased availability of crude, especially for processing at Venezuela's refineries. In future associations, not only should the formation of new mixed companies be contemplated, but thought should be given to the acquisition by CVP, or by other state entities and even private Venezuelan investors, of part of the capital stock in companies operating abroad in various petroleum activities. Eventually, and providing circumstances are propitious, total shares of such companies might be purchased, if at all feasible.

Although the policy and situation that have prevailed in Venezuela throughout the last twelve or thirteen years may now undergo a complete change, leading to intense activity involving the exploration of new areas and incorporating a great volume of reserves, we must not forget that in a not too distant future the time will inevitably arrive when Venezuela's petroleum resources will become exhausted. However, if Venezuela is able to secure sources of crude in other countries, it will be able to import it and continue using its own refineries and other petroleum installations. This might seem utopic, but it is precisely the tactic adopted by the United States in 1948, when it became a net petroleum importer, and also in the past by some Western European countries, such as England and Holland. These great colonial nations, after losing their overseas territories, were able to continue developing, due, partially, to the income they continued receiving from abroad in dividends, derived mainly from petroleum.

Venezuela could, therefore, progressively plan and carry out a program among others, of petroleum investments abroad, which would be a sort of security against the time in which its deposits under exploitation, and those to be discovered and developed, finally became depleted.

Participation by Venezuelan Entrepreneurs

I wish to emphasize the tremendous importance of the role that local entrepreneurs could, and must, play in this new phase that is beginning in Venezuela through the nationalization of its petroleum industry. We must not overlook the fact that on becoming nationalized, the petroleum industry passes from the private to the public sector. As a consequence thereof, the former will become weakened, and the public sector, or state, will become strengthened. This is, therefore, a powerful reason for intense and decisive action on behalf of Venezuelan entrepreneurs and professionals in the petroleum industry's private sector.

There exists in Venezuela a group of entrepreneurs which—although not too large—has become outstanding in Latin America due to its managerial capacity, efficiency, imagination, and reliability. This group is therefore a first-rate available resource, and it must be used in the petroleum industry to achieve the desired orientation.

As mentioned earlier, besides the undeniably political aspects of the petroleum industry—as a primary factor of the nation's resources and in orienting government action, externally more than internally—it is essentially a business. This commercial facet of the petroleum industry cannot be overlooked or sidestepped, due to the risk of precipitating it into an even greater decline. Consequently, now that the industry is to become thoroughly Venezuelan, many national entrepreneurs should contribute their efforts and managerial and commercial talents to the great endeavor of reactivating this extremely important facet and transforming Venezuela, I reiterate, into a truly petroleum nation.

On the one hand, activity by entrepreneurs must include direct participation and working with the government in handling state petroleum companies. On the other hand, it must include promoting and administrating oil service companies, both in the national and international spheres. It is therefore necessary to rapidly create a consciousness among Venezuelan entrepreneurs of the fact that the country needs them, that their cooperation is essential for achieving the high goals that have been set. Besides opening an important field to their own advantage, if they know how to wisely utilize the opportunities arising from the present state of affairs, they shall certainly be able to expand their range of essentially commercial activities.

DEVELOPMENT OF NEW AREAS

At the end of the 1960s, a great majority of oil experts were still predicting an abundant world supply of fossil fuels—referring particularly to petroleum and natural gas. Based on these forecasts, they asserted that petroleum-producing countries, especially OPEC members, had no solid reasons for demanding substantial price increases. However, opinions of others differed. Aside from leaders of some OPEC countries, Venezuela among them, independent United States petroleum producers also felt that as a result of the mandatory program covering import restrictions, lack of incentive was leading to a decline in production of the United States petroleum industry.

By 1970, the first signs of the approaching energy crisis became clearly defined. However, delays in developing new potential petroleum and alternative sources of energy prevented any solution to the situation. In 1971, 1972, and most especially toward the end of 1973 and beginning of 1974, the critical situation of energy supplies prevailing in the United States erupted. Other Western industrial nations were also facing the same problem, aggravated by the fact that most of them lacked domestic supplies, and increased consumption in the United States would obligate prorating the petroleum entering international channels. This situation seriously restricted the uninterrupted progress necessary to developed nations for maintaining the privileged position they enjoyed in all economic, technological, social, and political aspects.

These perspectives, plus other factors, such as the extraordinary price hike and the embargo imposed by Arab nations at the end of 1973, obliged world powers to anxiously seek new petroleum sources wherever they might be, roving from Ecuador up to the Siberian steppes, and to accelerated technological investigations aimed at taking advantage of the most diversified types of energy sources. In the oil world, the industry avidly attempted expanding supplies by exploring areas where prevailing conditions are increasingly hostile, such as in the Arctic and in open seas, or by conditioning hydrocarbons heretofore considered nonmarketable, such as those derived from lutites and bituminous sands.

Undoubtedly, the logical rise in prices favors the development of new areas and new sources of energy, and in view of this situation I believe that Venezuela, still possessing nonexploited areas of conventional and nonconventional petroleum, such as the vast Orinoco Tar Strip, should act swiftly and efficiently in this direction.

Sources of Conventional Petroleum

The Continental Shelf

Timid counsel never fails to bring unfortunate consequences. SIMÓN BOLÍVAR
Letter to J. M. del Castillo,
Bucaramanga, May 15, 1828.

Progress in open-sea drilling and production techniques has allowed the petroleum industry to penetrate continental shelves[4] in search of oil.

Venezuela has a coastline exceeding 1,510 miles, and hence possesses a

---

[4]A continental shelf is the portion of submerged land surrounding each continent. Due to width variations, its limit has been established as the point where the waters reach a depth of 100 fathoms.

vast continental shelf, although its width is, of course, greatly variable. Facing Boca Grande (Orinoco Delta), it penetrates approximately 65 miles into the Atlantic; in the east (Paria Gulf), it extends beyond the island of Tobago; facing the western coasts (Falcón state), it extends beyond Aruba; whereas facing the central coasts (between Aragua state and the Federal District), it is barely a half mile wide. It is quite possible that Venezuela's continental shelf contains petroleum in the areas shown in Table 10.1.

Bases for a vigorous exploratory program in this zone   Great investments are required for estimating possible resources in the continental shelf, because besides the high normal economic risks involved in a search for petroleum, costs of the most refined techniques must be added. Drilling is performed in open-sea beds, and it is necessary to foresee possible effects caused by marine currents, surfs, and storms. Two other important factors are the workers' welfare and safety and the need to adopt all measures for preventing ecological damage resulting from accidents, which are always possible in any exploratory phase.

To be able to add important volumes of hydrocarbons derived from the continental shelf to Venezuela's proven reserves, a systematic and ambitious exploratory program would be necessary. Governmental efforts up to the present have proved to be weak compared to the magnitude of work involved. This work should be carried out immediately in order to evaluate the potential of the entire shelf.

The Gulf of Venezuela

First the fatherland that formed our being; our
lives are nothing but the essence of our country.
SIMÓN BOLÍVAR
Letter to General Andrés de Santa Cruz, Popayán,
October 26, 1826.

Although the sedimentary basin of the Gulf of Venezuela is not under exploitation, I am referring to it separately because there are historical, political, and economic problems between Venezuela and its neighbor, Colombia, involving this region, which necessarily · must be fully explained.

From a strictly petroleum viewpoint, up to 1959 the Gulf of Venezuela was considered as part of the great Maracaibo-Falcón Basin. Thereafter, and in a new appraisal of sedimentary basins, it was classified as being independent. In the records of the IV Venezuelan Geological Congress, held in Caracas November 16–22, 1969, the sedimentary basin of the Gulf of Venezuela is described as follows:

> This great sedimentary basin, having an area of 15,300 square kilometers [5,900 square miles] almost entirely covered by the waters of the Gulf of

Table 10.1 **POTENTIAL PETROLEUM
IN THE CONTINENTAL SHELF**
(Thousands of acres)

Area	Surface
Gulf of Venezuela	3,780
Paraguaná-Guajira Shelf	2,200
Falcón Shelf	2,050
Cariaco Basin	6,920
Margarita-Northern Paria Shelf	4,770
Paria Basin	1,730
Delta Amacuro Shelf	7,388

Venezuela and the Coro Embayment, is the last "new frontier" in Venezuela promising petroliferous riches nearly as praiseworthy as those already known in the Basin of Maracaibo. However, the Gulf Basin still represents mere promises because not a single well has yet been drilled therein and, hence, the first barrel of crude has yet to be produced. Notwithstanding several seismic surveys performed in recent years, and a certain degree of geological continuity between the Gulf of Venezuela Basin and that of Maracaibo, the Gulf basin still holds many secrets concerning its stratigraphy and structure; its source beds, oil bearing and reservoir rocks; and the location and depths of its best treasures. Seven important dry exploratory wells were drilled on adjacent coasts, proving to be of great value for the geological interpretation of the Gulf. However, expectations of this basin's promise are based on the projection of known geological data in adjacent zones toward the Gulf, and on seismic reconnaissance surveys covering it almost entirely.

The Gulf of Venezuela—its western boundary formed by the Guajira Peninsula's eastern and northern coasts—historically and legally (as its name so indicates) belongs to Venezuela. Since Spain's political organization of its South American colonies, the Cape of La Vela, on the Antilles Sea and in the western extreme of that peninsula, had been the starting point of the frontier between the present republics of Colombia and Venezuela. This historic, legal, and political fact was accepted by both countries when they signed their first sovereign action, upon becoming independent from Spain in 1810, and was thus established between both countries in their first respective constitutions at that time.

Subsequently, the starting point of the border was transferred from the Cape of La Vela to the Castilletes Promontory, situated toward the east, whereby Colombia penetrated into Venezuelan territory, and the Guajira Peninsula on waters of the Gulf of Venezuela appeared as belonging to Colombia, according to the Limits Treaty signed between both countries in 1941. However, this treaty is contrary to the Constitution of Venezuela, which establishes that no part of national territory, under any concept whatsoever, may be granted or transferred to any foreign power. This Venezuelan constitutional ruling, which has remained intact to the present, bars Venezuela's President, or Congress, through special powers bestowed on them, from ratifying treaties signed with Colombia, because future treaties, marginal to Venezuela's constitutional rights, would still be invalid thereby.

Notwithstanding this historic and juridical fact, and even though public opinion in Venezuela, prior and subsequent to 1941, has opposed the 1941 treaty, Colombia—cognizant of the treaty's nullity petition— proposed to Venezuela the demarcation of marine and submarine waters in the Gulf of Venezuela. The objectives pursued by Colombia are quite clear. From a political and legal viewpoint, Colombia seeks to secure and reinforce its territorial appropriation from Venezuela (which lacks foundation in the light of both countries' national and international rights). From an economic viewpoint, Colombia's ambition for territorial expansion is now intensified by the petroliferous wealth undoubtedly contained in the Gulf of Venezuela, as evidenced by the discoveries of natural gas on the Guajira Peninsula.

It is quite possible that Colombia's secular expansionist policy might exasperate Venezuela's passivity, thereby originating conflictive situations in this part of the world. Should this be the case, it will be no fault of Venezuela, since it is Colombia that is refusing to honor its obligations, as described previously. In November 1974, Bogotá's press published a statement by retired Colombian militarists expressing their conviction that Colombia should not give (read "return") to Venezuela one single inch of territory, i.e., the military sector of Colombia was inciting nonfulfillment of the country's prevailing international obligations with Venezuela.

It is important therefore to stress the fact that the so-called rights being claimed by Colombia over marine and submarine areas of the Gulf of Venezuela may lack legal support because they are based only on the 1941 Limits Treaty. A petition to nullify the law that ratified the treaty is now before the Venezuelan Court of Justice. If granted, the effects of the treaty shall become null and void in Venezuela.

Sources of Nonconventional Petroleum

Under this heading I refer to a group of hydrocarbon deposits which, up to a few years ago, had no industrial application, because with existing available techniques, their exploitation and processing was nearly impossible, or in all events too costly. In effect, the description "nonconventional" applies to the methods of exploitation, because the hydrocarbons are crudes, like the others. The difference lies in their gravity, for which reason petroleum proceeding from these sources is more adequately

Table 10.2 **POTENTIAL PETROLEUM SOURCES**
(Billions of barrels)

Sources	Estimated volume
Bituminous lutites of Colorado, Wyoming, and Utah	1,740
Orinoco Tar Strip	700
Athabasca bituminous accumulations	710
Total	3,150

described as extra-heavy crude. Such petroleum, therefore, in general has very little fluidity and a low gravity (its specific weight is equal to or heavier that that of water). The most reknown accumulations are those located in Athabasca (Canada), Colorado, Wyoming, and Utah (United States), and the Orinoco Tar Strip (Venezuela), apart from similar formations existing in Brazil, Europe, and Asia.

The necessary technology for exploiting extra-heavy crudes varies according to the type and depth of their formations. The volume of petroleum existing in some of the known deposits possibly exceeds by fourfold proven world reserves (700 billion barrels, as of December 31, 1971), which explains the great interest at present in developing these potential petroleum sources. Table 10.2 shows some figures in this respect.

It is still premature to establish the volume of these petroleums that could be extracted, but as an indication I might mention that the recovery factor estimated for the Athabasca sands is on the order of 30 percent.[5] In the Orinoco Tar Strip it is estimated at around 10 percent. In general it is thought that the volume of petroleum these two deposits could yield would at least be equal to the present volume of proven reserves in the world.

Venezuela should intensify its present exploratory programs in the Strip in order to reach a more definite estimate of the reserves in this region and also of those to be developed, in a relatively near future, in this new and important petroleum horizon.

THE GAS INDUSTRY

World public attention is increasingly focusing on petroleum and on all events directly or indirectly related thereto. In Venezuela, one of the

---

[5]This is a high recovery factor, because these hydrocarbons are located at the surface. Open-sky exploitation is effected, as in open-pit mining.

principal petroleum-exporting countries, all world events relating to oil possibly have an even greater resonance in local public opinion. On the other hand, and through diverse communications media, a series of debates are taking place concerning matters related to aspects which— also directly or indirectly—are tied to Venezuela's oil industry proper.

In Venezuela, the imminent nationalization of its petroleum industry is, of course, the center of attention. Hence all aspects relating to this process—e.g., how it shall, in practice, take place; administrative and technical aspects involved; which companies might develop from nationalization—are subjects of intense actuality being discussed concurrently with international news concerning "petroleum diplomacy" or the "petroleum cold war."

Besides nationalization, other matters of interest connected with the petroleum industry are also under discussion: e.g., the creation of an oil fleet; the development of a naval industry; and the national petrochemical industry. However, a subject at present absent from Venezuelan public analysis is the industrialization of natural gas.

For many years natural petroleum gas was not used by concessionary companies; rather, it was wasted by burning. This happened in nearly all cases up to 1945; since then, it has been used, but not in industrial processes proper. Instead, it has slowly and gradually been reinjected into the wells, in the secondary recovery process.

During President Rafael Caldera's government (1969–1974), private foreign companies submitted gas liquefaction projects for consideration. Toward the end of 1973 the government presented Congress with a project for execution by the state. This project—which fundamentally consisted of the construction of a plant having a capacity for processing approximately 20 million cubic meters per day, with an investment of approximately Bs2.3 billion, including acquisition of three methane tankers—was at that time not considered by Congress. Carlos Andrés Pérez's present government, which took office in March 1974, has not revived this project, nor has the project been subject to modifications.

The government is undoubtedly taking into account the fact that Venezuela might possibly need its natural petroleum gas, and also the free natural gas apparently existing in vast quantities in the eastern area of the country. In this respect, future studies must of course contemplate the great quantity of gas that will be required by the iron industry, which at present is being actively expanded. In fact, SIDOR (the state siderurgical company), today producing more than 1,500,000 metric tons of steel per year, is expanding to meet a projected installed capacity of more than 4,800,000 metric tons per year by 1978. On the other hand, two great companies for producing various types of steel are being promoted; they will probably be of mixed capital and have a capacity of approximately

5,000,000 metric tons each, per year. These projects will be executed in the Guayana zone, where rich iron ore deposits exist. Thus, toward the mid-1980s, Venezuela may have an installed steel production capacity of approximately 15,000,000 metric tons per year, and part of this production—through the prereduction process—will require increasingly higher volumes of natural gas.

Furthermore, at El Tablazo, near Maracaibo, Zulia state, in the western area of the country, Venezuela has the necessary bases for developing an important petrochemical industry. If the great administrative and technical failures traditionally characterizing IVP (Venezuelan Petrochemical Institute) are overcome, requirements of gas proceeding from the Maracaibo Lake (fundamentally petroleum gas) by this petrochemical industry will also be voluminous.

It is for these reasons that the Venezuelan state should immediately make a complete study for estimating the country's gas reserves and requirements in the forthcoming ten or twenty years. To that effect, the Ministry of Mines and Hydrocarbons, in its annual reports, has been submitting information concerning reserves of natural gas, but the figures indicated correspond, at least by 5 percent, to free natural gas and, by more than 95 percent, to natural petroleum gas. This relation shows that Venezuela does not thoroughly know what its real natural gas resources are. According to official statistics, the country has more than one billion cubic meters of natural gas reserves, mostly corresponding to oil gas, i.e., the natural gas associated with the production of crude. This proportion of Venezuela's gas reserves is therefore directly related to the rate of crude extraction, and its proportion, in relation to petroleum, tends to increase while deposits decline. Thus it will decrease while production—for preservation reasons—is reduced.

Apparently rich deposits of free natural gas exist in the western area of Venezuela. However, the law that reserves the gas industry to the state, approved by Congress in August 1971, prohibits the exploitation of this resource. Once the country has acquired a thorough knowledge of its future needs, Congress may have to reform this law in order to permit a rational exploitation of free natural gas. The thorough investigations that Venezuela must make—as soon as possible—concerning natural gas reserves and volumes that will be required for national development on the short, medium, and long terms (besides other factors) will allow a conclusion as to whether it would be recommendable or not to install a gas liquefying industry in the country. Such an industry requires major investments and local infrastructure jobs, for installing the plants and for the necessary means of transportation for exporting the gas (methane tankers).

It is quite possible, however, that the conclusion will prove to be negative, i.e., that it will not be recommendable to install a gas liquefying industry for exporting surplus liquefied gas. Besides other reasons, such as preserving this energy resource because it is scarce in relative terms and ecologically pure, it could be that Venezuela does not have sufficient reserves to justify an export industry of the product in its primary state. Or, it might also mean—which would be good news—that the country will be able to develop petrochemical and siderurgical industrial projects, among others, of such volume that would permit local processing of this natural resource for domestic consumption and for exporting finished products, derived from natural gas, on a large scale.

THE PETROCHEMICAL
INDUSTRY

To do things right they must be done twice: The first time teaches the second.   SIMÓN BOLÍVAR

Letter to General Antonio José de Sucre,
Guayaquil, May 24, 1823.

Venezuela is a country having ideal conditions for the existence of a booming petrochemical industry, but it does not have such an industry. Although Venezuela's petroleum industry achieved world predominance in 1928, it was only toward the end of the 1960s that it started executing the planning and development of a true petrochemical industry. Venezuela has a very unsatisfactory history in this respect, which must be improved through creative, sustained, and tenacious diligence.

In 1954 Venezuela's Department of Economics of the Ministry of Mines and Hydrocarbons drew up a project for creating a national petrochemical industry, including the following aspects:

- Administration
- Installation and production
- Supply of raw materials
- Product experimentation and consumption
- Personnel training

For the purpose of carrying out this project, the Venezuelan Petrochemical Institute (IVP) was created in June 1956, as an autonomous institute attached to the Ministry of Mines and Hydrocarbons and having its own legal and patrimonial capacity. Article 3 of its constitutional statutes states: "The objectives of the Venezuelan Petrochemical Institute shall be to study, establish, operate and develop industries for the purpose of exploiting minerals, hydrocarbons and any other product related to the petrochemical industry." Unfortunately, up to 1974 these objectives had not yet been attained.

Under this article, the Institute is empowered to create and operate industrial establishments. In other words, all matters related to the petrochemical industry in Venezuela are the responsibility of IVP not only directly, but also through companies that it might establish; it is further empowered to form industries in order to exploit other minerals, and thus the so-called satellite industries or companies and mixed companies in which the Institute participates were created.

IVP and the mixed companies created (or to be created) are in fact fit tools for developing a petrochemical industry in Venezuela; however, the experience accumulated since 1956 reveals that achievements to the present have not proved to be satisfactory. The Venezuelan state's policy and administration in this field have been so erroneous that its performance might, in general, be classified as catastrophic. According to the General Comptrollership of the Republic, the organization in charge of supervising and investigating the state's resources, the government's budgeted assignment to IVP up to December 31, 1971, amounted to Bs1.3 billion ($300 million), and according to the same source, accrued losses to that date exceeded Bs800 million ($184 million), thereby decreasing its patrimony to approximately Bs550 million ($126.5 million).

But apparently the situation is even worse. Besides innumerable objections over the serious administrative, accounting, and investment failures, among others, the General Comptrollership reported to Congress that "the General Balance Sheet and Profit and Loss Statement of IVP do not reasonably reflect the financial status of the institute as of December 31, 1971, nor the results of its operations during the exercise then ending."[6]

I cannot analyze here the many aspects related to these exorbitant losses, to the administration of the Institute, to the regrettable state of IVP's installations at Morón (Carabobo state), or to the excessive investments made by some of the Institute's companies, such as Nitroven, created for producing ammonia and urea. Nor can I go into the manipulations and operations that occurred in connection with the creation and development of that company, or of any other. I can only mention the undeniably negative evidence of the voluminous losses incurred by IVP, which, regardless of Venezuela's prominent position for many years among petroleum-producing and petroleum-exporting countries, up to recently was a petrochemical institute in name only.

Finally, in the petrochemical program initiated in the mid-1960s— which has gained force in recent years—the creation of mixed companies

[6]General Comptrollership of the Republic, *Report to Congress, 1972*, Caracas, 1973.

in which foreign capital will participate (in addition to Venezuelan private capital, wherever feasible), possibly contributing technology and external markets, is considered one of the most important tools for development. This alternative does seem plausible; however, the Venezuelan state (IVP), and its budgetary bureaus, as well as, Congress, must be extremely careful in making sure that associated foreign companies make truly advantageous contributions to the country in the aforementioned technological and marketing aspects (and in other fields, such as finance). They must be alert against "contributions" by foreign partners who represent only a minimum investment of capital through sales of patents and certain processes, or a minimum investment of machinery and services, at excessive prices, which are obsolete by the time they are provided or sold to new mixed companies promoted by IVP. We must not forget that, for these or similar reasons, the constitution of mixed companies and contracts drawn up with them have, in some cases, been questioned by public opinion and in Congress.

Notwithstanding the aforementioned negative aspects, the state has already demonstrated a certain degree of administrative capacity in operating official agencies in other fields. I therefore do not see why it should not be able to develop and even improve its administrative capacity in the petroleum industry.

THE PETROLEUM FLEET

The time has thus arrived for us to take out to sea.
SIMÓN BOLÍVAR
Letter to Mariano Montilla and José Padilla,
Caracas, January 27, 1827.

It is well known that up to now exploration, production, refining, and marketing of Venezuelan petroleum has largely been in the hands of concessionary companies. Likewise, sea transportation of the voluminous flow of crude and by-products exported (more than one billion barrels per year) is also handled by the concessionaires, through their affiliated companies, and, although in a lower proportion, by independent shipowners. Furthermore, coastwise shipping has been carried out by these affiliated companies.

Official estimates place total sea transportation, to or from Venezuela (imports and exports of all kinds of products), at the extraordinarily high level of 10 percent of the world total. This great volume of business—in which hydrocarbon exports predominate—has been monopolized by foreign shipowners in an overwhelming proportion. Therefore, where petroleum transportation is concerned, Venezuela again shows lack of initiative by its government and private sector in what should already be a great source of employment, investment, revenue, income, and economy in general, as well as another important tool in Venezuela's petroleum policy.

Many speculations have been made in Venezuela concerning the possibility of building ships locally, especially tankers, and of local ship repairing on a large scale, but these subjects so far have only been part of Venezuela's underdeveloped ideology where its petroleum industry is concerned. Granted, some comments and opinions have been expressed concerning the importance for Venezuela to have its own oil fleet, and some Venezuelans have even outlined certain projects in this connection. However, oil-company spokesmen and certain national sectors, especially entrepreneurs, tenaciously alleged, until quite recently, that Venezuela had no marine tradition or vocation whatsoever and furthermore lacked any nautical preparation or experience. They asserted that, since Venezuela's public sector had irrefutably demonstrated its remarkable inefficiency in nearly all the fields in which it had participated, should it enter this extremely complex field without the necessary experience and preparation, it would definitely contribute to increasing the costs of Venezuelan fuel in world markets and thus further debilitate the already fragile competitive position of its petroleum industry.

These facts, alleged by opposers to the creation of a national petroleum fleet, are quite true. The country does not possess the tradition, preparation, or experience necessary for immediately handling, without outside cooperation, an important petroleum fleet of its own; neither have Venezuelans ever shown—since the advent of the republic—any vocation for seafaring activities or for the nautical industry. It is also as true as the light of day that Venezuelan governments have always been notable for their remarkable inefficiency.

I feel, however, that after more than fifty years of petroleum exploitation by foreigners, who thus created in Venezuela one of the greatest oil industries in the world, Venezuelans cannot possibly continue adopting a pessimistic, resigned, and nearly fatalistic attitude. They cannot remain inert, with a passivity and indifference bordering on lack of patriotism, placidly watching the unending parade of tankers on the Maracaibo Lake and along the vast coasts of their country bearing exotic flags and insatiably carrying overseas the treasured black gold extracted from their country's ground for feeding the machinery of industrial nations. This fundamental need to awaken Venezuela from its inertia, in this case where sea transportation is concerned, in order to become a *truly petroleum nation,* is in itself a sufficiently powerful reason for facing all other objections, however important they may be. Venezuela should therefore develop its own fleet and demand that its exports of crude and by-products be transported in Venezuelan tankers. We must realize that some Eastern and African petroleum countries are developing projects aimed at having their own fleets, and six have already achieved that goal.

In July 1973, Venezuela's Congress passed the Law of Protection and Development of the National Merchant Marine, which includes the legislative basis needed by the country for starting its own petroleum fleet, among other aims. Another aspect to be borne in mind is that in order to carry out this vast project one of the essential requisites is the installation of deep-water terminals able to receive modern supertankers. These giant vessels have a deadweight capacity of close to 500,000 metric tons (the *Globtik Tokyo*, for example, has a deadweight of 483,000 metric tons). Because of this modern characteristic of world transportation of petroleum, it is regrettable to see that dredging services of the Maracaibo Lake access channels are at present only able to accept vessels having a maximum 40-foot draft, i.e., tankers of approximately 75,000 metric tons deadweight. Comparing this capacity to that required by the great seafaring tankers, or even with modest 100,000-metric-ton oil tankers, we are able to see the urgent need for Venezuela's government to execute a plan to increase the terminal capacities for receiving petroleum tankers. At the time of this writing it is, in fact, advancing concrete projects for carrying out a very ambitious program in this field.

Venezuela has obviously not yet developed its ship-repairing industry, which would absorb national costs and generate sources of work, both directly and through the companies that nourish this type of industry. It has, furthermore, not taken any important step—truly nationalistic and closely tied to ship repairing—toward developing a shipbuilding industry capable of building its own future oil fleet, and thus toward eventually becoming independent in this field. The country has again lost valuable time here; but I must emphasize that there is still time to become the owner of an important shipping industry, and to that effect it is necessary to adopt measures aimed at building its own tankers, by means of a thoroughly studied and precise program contemplating adequate technical association, among other aspects.

In addition to a basic petroleum policy and its captive market, Venezuela's position as a great iron and steel producer must be especially taken into account, as well as its cheap energy, since they are elements of fundamental importance for developing a shipping industry. In the near future SIDOR's new laminating plant will launch products essential to shipyards on the market. Near the siderurgical plants in the eastern area of the country is the great Guri hydroelectric complex (Bolívar state), which is a million-dollar generator of cheap energy. Finally, Venezuela covers a privileged geographical position in the northern part of South America, with an extensive coastline.

Another aspect that must not be overlooked is that shipyards represent a vast economic investment. However, Venezuela possesses the necessary funds by not having yet generated a sufficient number of new industrial projects, which is not at all logical considering the resources it has available. Venezuela should also take into account the need to limit acquisitions from abroad as far as possible. This would be quite feasible upon establishing its own petroleum fleet, since it thereby should be able to avoid spending an important proportion of its funds in having to purchase tankers. It should in fact even take advantage of this state of affairs for creating its own shipping industry and use to the utmost its funds for importing only absolutely essential elements.

Venezuela's government is developing important studies for installing a shipping industry, in which the need for three shipyards is mentioned: one for building petroleum tankers (60,000 to 80,000 metric tons deadweight); another for ships of 12,000 to 15,000 metric tons deadweight for general freight or petroleum derivatives ("clean cargo"); and a third destined for repairs and transformation of vessels having up to 80,000 metric tons deadweight. The investment necessary for these shipyards has been calculated at approximately $240 million, and it is estimated that employment for more than 5,000 persons, including professionals, technicians, and workers, could be generated.

Since such a vast project as shipbuilding requires many years to become factually established (estimated by Venezuela's government at five years), feasibility studies should therefore be rapidly put into execution. In the meantime, a petroleum fleet should be gradually created, and experience should be accumulated on the basis of ship-leasing programs or various types of associations with foreign companies, within the margins permitted by law. Projects such as these bring to mind the fact that quite frequently in Venezuela ideas and economic development plans have been drawn up and never executed, even though resources and favorable conditions prevail. It is for these reasons that I believe the time has definitely arrived for these important plans and constructive ideas to cease being theoretical and to be put into practice.

To those individuals who believe that Venezuela is not yet prepared for penetrating into this industry, I might suggest they look at Brazil—which at present owns an important shipping industry, even though it is not as geographically favored as Venezuela and does not have a captive domestic market of the extraordinary magnitude of Venezuela's. Nor must we forget that up to fifteen years ago Spain's position in the shipping world was insignificant, whereas today it holds third place.

With its vast financial and basic human resources, Venezuela should be able to develop activities such as those described in this chapter to become a truly petroleum nation. An awareness is already becoming evident in public opinion concerning the need to achieve these goals and to over-come passivity and serious public administrative inefficiency. A clear national consciousness must propel the nation's leaders toward develop-ing positive and audacious action within this field.

ELEVEN

Goals and Strategy within the World Order

INTRODUCTION

It is my intention in this last chapter to contemplate some fundamental goals for Venezuela and other oil-exporting countries within the changing world order, and consider certain strategies for achieving those goals.

A priori, these objectives, and the determination of the strategy to be followed, are fully justified for several reasons:

First, Venezuela's destiny, as well as that of all other petroleum countries, is closely tied to its oil industry.

Second, the OPEC countries are targets of certain threats expressed in various forms by great petroleum-consumer nations, they should endeavor to unify their petroleum policy as far as possible. The individual petroleum policy of each country should be oriented within the general context of the group, as provided for in the Baghdad Agreement that originated OPEC.

Further, although the petroleum industry has always been characterized by its eminently international status, its importance is now enhanced because it has become one of the most important individual factors in the development of world economy. The industry is today under the direct, or indirect, control of the countries that possess the raw material.

We can thus see that petroleum-producing and petroleum-exporting countries must assume the dual role of defending their common interests and of orienting and offering their effective support to other Third World countries in their struggle to achieve an improved equilibrium in world economy and world distribution of wealth.

The countries richer in petroleum have long been among the more underdeveloped on earth. The new reality now gaining force is that voluminous revenue is being transferred to this small group of countries from the richer on earth. This fact should not only contribute to the rapid development—at all levels—of petroleum-exporting countries, but even lead to establishing a more equitable equilibrium of world wealth. In this respect we must not forget the anguishing phenomenon as stated by the Organization for Economic Cooperation and Development (OECD)[1] in its Annual Report on Development Assistance Activities of 1973: that while the economy of the world's richer nations increased by 60 percent in real terms (i.e., deducting the effects of inflation) in the 1960s, their assistance to underdeveloped countries, many so impoverished that their inhabitants die of starvation, decreased by 11 percent.

It is therefore imperative today for petroleum-exporting countries, in addition to becoming developed, to contribute through their economic power to the establishment, at long last, of a more balanced world equilibrium. This economic power is placing petroleum-exporting countries in the role of obliged leaders of Third World countries in their effort to overcome their economic dependency on industrial nations.

Prevailing facts that must be taken into consideration by petroleum countries when adopting strategies and aims for future lines of action are: the need to cooperate with underdeveloped countries lacking petroleum, especially the most impoverished; the common causes that unify all underdeveloped countries in their struggle for economic independence and for attaining fairer trade relations with industrialized powers; the industrial nations' dependence on petroleum and other raw materials and natural resources produced by underdeveloped countries; and the phenomenon of interdependency, becoming increasingly accentuated among developed nations.

Within the group of petroleum-exporting countries, Venezuela must continue in a leading role because of its past prominent position as a principal petroleum exporter and its major advancement compared to other export countries, suggesting how to handle their relations with oil companies and import countries, as I mentioned previously.

This experience of Venezuela, and its relatively greater development in economic, political, and social affairs (compared to other OPEC coun-

---

[1] As indicated by its name, the main purpose of this organization is to promote cooperation and development among its members; it was created in 1961 as a substitute for the Organization for European Economic Cooperation (OEEC). Its members are West Germany, Austria, Belgium, Canada, Denmark, Spain, United States, France, Greece, Holland, Ireland, Iceland, Italy, Japan, Luxembourg, Norway, Portugal, United Kingdom, Sweden, Switzerland, and Turkey.

tries) make its future participation in the petroleum world important for all other oil-exporting countries, even though its reserves and production are markedly inferior to theirs.

In the following pages I shall refer in more detail to some aspects of the world state of affairs, fundamental to my observations concerning the goals and strategies that should be pursued by Venezuela in becoming a truly petroleum nation.

WORLD STATE OF AFFAIRS AND TENDENCIES

The Dependency of Underdeveloped Countries

The most equitable judgements are the least appreciated.   SIMÓN BOLÍVAR

Letter to General Antonio José de Sucre,
Arequipa, May 15, 1825.

The Process of Dependency

The development of the small group of highly industrialized nations—which commenced centuries ago—is in great proportion the fruit of their inhabitants' toil, sacrifice, national ethics, discipline, patriotism, foresight, and unrelenting willpower. However, alongside these positive aspects lies the fact that their progress was achieved partially at the expense of poor countries—more numerous and housing today more than two-thirds of humanity—which has shaped over the centuries the grave phenomenon described today as **dependency.**

Within the frame of this dependency by Third World countries on developed nations are found the following fundamental elements: the basically monoproducing status of underdeveloped countries; the sale of their raw materials, frequently at depressed prices, to industrial nations, although they purchase from the latter finished products at prices increasingly higher, thereby contributing each year to further widening the gap separating the two groups; the technological dependency and intellectual submission of underdeveloped countries; the massive consumption and continuously higher standard of living in industrialized nations, causing an unceasing rise in the cost of their manufactured goods, and in general, inflation in their economies; the direct foreign investment in the agricultural and mining sectors of underdeveloped countries; the acquisition, especially after World War II, by multinational corporations based in industrial nations, of the principal manufacturing companies operating in the underdeveloped countries.

The foregoing factors powerfully contributed to structuring the economic process within which some nations have achieved development, or industrialization, and others have not. This reality has contributed to intensifying the growing discontent in dependent countries, which has crystallized into an acute reaction of ingrained nationalism. In my opinion, an accurate diagnosis of this process requires a dispassionate analysis, by both industrialized and underdeveloped countries, shorn of preconceived doctrines of self-interest.

On the basis of these outlines, let us briefly review how the material and spiritual relationship of dependency originated between the developed and underdeveloped worlds.

First, I would mention that there exists a tendency in underdeveloped countries, influenced by concepts implanted by industrial nations, to believe that underdevelopment—and also development—is but a moment in the economic, political, and cultural evolution of a nation, as if the world were an aggregate of isolated countries whose degree of development (placing them within a general classification of rich or poor) depended exclusively on their individual historical performances without any interconnection with each other. This is certainly a naïve view, because the process of development cannot be considered as a competition that started at a given place, and at a specific moment, prior to the industrial revolution. It cannot be said that, as in automobile races, some countries, because of their own and exclusive efforts, markedly progressed, while others lagged behind. This assertion has no validity, because the formation of a capitalistic world permitted the creation of an international system of dependency that placed world economy under the decisive influence of a group of European countries (including Russia), to which the United States became incorporated in the nineteenth century and Japan in the twentieth. The remaining poor and underdeveloped countries became dependent on that small group of advanced nations, which, nourished by the former's cheap raw materials and even by their markets, rapidly progressed.

This material dependency became cemented within the prevailing international economic order. Industrialized nations as a rule acquire raw materials from underdeveloped countries at depressed prices and then sell their own finished products at high prices. Because of this trade relationship, the higher benefits obtained by industrial nations are logically transformed into a higher concentration of capital and into a higher standard of living for their inhabitants, in contrast to the lack of funds and tremendous social problems confronted by countries with hardly any industrial development.

This unequal trade relationship has contributed to carrying industrial nations to a privileged superior position, consequently allowing them to exercise predominance over the economies of the underdeveloped countries. In other words, the economic life of underdeveloped countries depends in practice on the developed nations.

Such a status of dependency is not a phenomenon exclusive to our times. It is a characteristic constant in the history of peoples, existing throughout various forms of social and political organization. Tribes, empires, states, and nations have been dominant or dependent throughout the centuries, according to whether or not they possess the economic and technological resources in a given stage of their evolution.

This prevailing order tends to remain inflexible in that developed nations continue holding a concentration of power mightier than that held by underdeveloped countries. As long as the trading relationship between both worlds remains unchanged, the gap separating them will persist, with its load of imbalance and tension of confrontation. Sooner or later this tension must lead to drastic transformations.

Nevertheless, it is by far easier to change economic laws than physical laws. Joint action by Third World countries would be able to partially or totally break the system. Since this situation is the result of a correlation of forces, a combination of power would suffice to cause a substantial variation in the policies of countries producing raw materials and to change the prevailing system and the relationships between the two worlds, causing the one to become less dominant and the other less dependent.

But these changes in the balance of forces will not be attained by expecting developed nations to extend to the Third World countries a fairer treatment in their trade relations. In all walks of life, the psychology of the dominant person does not incline toward deeds of cooperation that he might deem, in principle, contrary to his own interests. It will therefore fall to the underdeveloped countries to take the initiative for restructuring their relationship with industrial nations. To achieve this, they must rely on their own forces and their own resources, and they must be driven by their need to overcome their inadequate development and to emerge from the position in which they now find themselves.

The material dependency of Third World countries, therefore, has root in the survival of the present system of international trade between industrial nations and economically underdeveloped countries. The latter are gradually becoming aware of their power derived from possessing the raw materials essential to the functioning of those industrial nations. Therefore it is hoped that, over the short or medium term, the future will bring important changes.

Certain spiritual elements have also conditioned certain peoples to accept or reject dependency.

The conquest and colonization of some peoples by others becomes consolidated when the subjugated parties assimilate or accept the religion, customs, and in general, the culture of the conquerors or colonizers. Such ties have favored the economic exploitation of subjugated peoples, because a state of spiritual dependency is formed that frequently persists even after political independence is attained.

There are numerous cases where colonies have, on becoming independent, remained tied to the economic influence of their mother country because their social and cultural structures have not permitted them to overcome their traditional ties to dependency. However, history also reveals how other countries, having endured exploitation by imperial and colonial powers, have not forgotten those days of exploitation, and how, out of their humiliation, a spirit of rejection and resistence prevails against any new processes of penetration and dependency.

In recent history, Japan and China are fine examples of such a reaction and are worthy of mention here because both are located, respectively, within the ideological sphere of imperialistic nations, capitalist and socialist. Upon closing their doors to any foreign imposition that might gravitate upon their internal policies and spiritual life, they revealed themselves to be the possessors of genuinely nationalistic qualities, which became translated into a rejection of possible future relationships of dependency. They adopted, as far as they have been able to maintain it, an affirmative position of spiritual independency.

In this modern era we can clearly see how this spiritual process occurs, culminating in the dependency of countries today classified as the Third World. It is a process that has contributed to the development and prosperity of the small group of industrial nations today ruling the destinies of the world. Within this train of thought, let us briefly glance at some of the more important events leading to Third World dependency.

It is well known that the great geographical discoveries by Iberian navigators (Spaniards and Portuguese) were basic to the evolution and economic development of Western Europe. In a way, Western Europe was a peninsula of the Asiatic Continent, surrounded by Arabs and Turks. This situation became aggravated with the conquest of Constantinople by the Turks in 1453, and the consequent final collapse of the Byzantine Empire. When Crimea succumbed to Turkey, the Black Sea became an internal Turkish lake, with the conquests of the Ottoman Empire extending up to Asia Minor and the Balkans.

With the discovery of America in 1492, a navigational route toward the Western Indies was opened, enabling Europe to emerge from its secular stage of medieval underdevelopment and enter into a precapitalist period. Pearls and gold started to flow from America, and spices from Asia. The Turks were defeated in Corinth, and the Portuguese and Spanish fleets started trafficking on a worldwide scale, leaving Asia, Africa, and America open to conquest and expansion by Europeans and permitting the military and economic expansion of Europe and its entrance into a mercantile era, which marked the end of feudalism and cleared the way for capitalism.

In the colonial era of the Iberian-American world, and for the purpose of extracting precious metals and obtaining tropical products needed for their homelands, the European powers interfered in the settlers' existing relations, reorganizing their local economies on the basis of slavery and other forms of forced labor. This created the agrarian and institutional structures that, in certain forms, have survived in western underdeveloped countries up to the present time.

Subsequently, in the nineteenth century and in the first half of the twentieth, the industrial revolution—which began in the eighteenth century in Europe and followed in the United States—established a world economic system characterized by intense investment by those countries for promoting the production of food and raw materials in the rest of the world, while they themselves became specialists in manufacturing and finance.[2]

The aforementioned process leads to a fundamental aspect which, notwithstanding the changes that have occurred, especially political, has remained essentially sound up to the present. I refer to what we might call the dialectic unit "development-underdevelopment," according to which development presupposes underdevelopment and vice versa; i.e., the existence of the opposite position as a contingent condition. This dialectic unit commences in our modern era with the geographic discoveries and European colonization of several continents, originating a complementary world of poor countries, producers of raw materials, on the one hand, and of industrial nations, consumers and transformers of those raw materials, on the other.

Along with its colonial inheritance, and as a consequence of the process of dependency,[3] Latin America became specialized in exports of primary products in general. This, in turn, determined another series of social, economic, and political structures and institutions, as well as the development and strengthening of the privileged dominant classes.

Subsequently, in the twentieth century, the two world wars (especially World War II), and also the Great Depression of the 1930s, caused important changes in the prevailing international system of economic

---

[2]This process is well summarized in Jacques Pirenne, "The Great Currents of History," *Universal History*, Editorial Exito, Barcelona, 1961, 8 vols.

[3]See Mariano Picón Salas, *Dependency and Interdependency in the Hispano-American History*, Ediciones Cruz del Sur, Caracas, 1962.

relations. In Latin America, an era of industrial development began, based on the substitution of imports. Especially in the larger countries of the region, and since the 1950s, an important manufacturing sector developed, comprising a local professional managerial class and group of technicians, as well as an incipient industrial proletariat. The process included the formation of a series of financial, marketing, and educational institutions, governmental and private.

This industrial development, which theoretically should have led to debilitating the dependency of the underdeveloped countries, contrarily led to increasing it. Essentially, the original plan was based on the hope that wide protection and stimulation of local industrial activities, plus other supporting measures adopted by governments, would lead to a rapid industrial development. This, it was thought, would lead to the modernization and advancement of other sectors within the economies. In turn, this process would improve the people's social conditions, following the pattern of the industrial revolution in Western Europe and the United States. However, certain factors were overlooked.

In the first place, a great proportion of the profits derived from that industrialization were transferred abroad as payments for capital assets, with increasingly higher costs, and as transfers covering royalties, insurance, transportation, financing, and so on.

In the second place, although the process of import substitution between the 1940s and 1960s stimulated the development of national industrial sectors and managerial classes, these sectors were subsequently absorbed in high proportion by the subsidiaries of foreign companies. This caused a great part of the benefits expected from industrial development to be remitted abroad as profit transfers. It further eroded the managerial class, which lost strength and initiative on becoming, in high proportion, absorbed or replaced through the employment of foreign administrators by the new companies controlled from abroad. Finally, large amounts of national savings were allocated to the financing of foreign companies, limiting opportunities for national entrepreneurs and decreasing the flow of currency to these underdeveloped countries.

This process of acquisition of national companies by foreign entities, which weakened the bases for local enterprise, contributed to increasing the concentration of national capital within the small remaining group of industries that were not controlled from abroad. Likewise, differences became accentuated between rural income and that of the cities, where the increasingly mechanized industries became concentrated and gradually replaced the traditional local rural activity with its high labor employment. This industry, limited to the highly protected national markets, has been unable to adequately confront spiraling unemployment and underemployment (aggravated by the high rates of demographic growth), which in many underdeveloped countries reaches 25 percent of total available labor.

The tendency today, therefore, is to study Third World countries' development not in an isolated manner, but rather as part of the international capitalistic system, the dynamics of which have a pronounced influence on local processes.

Other factors have combined with this process of economic domination by the small group of industrial nations. For example, these nations developed, especially since World War II, a strategy aimed at exercising control over the various international organizations of a political, financial, and social nature. This permitted them to consolidate their dominant position over world economy. It is known that the United Nations Security Council has been dominated by the great powers, both Western and socialist, which have the strength of veto in its decisions. It is only recently, and through the incorporation of a large group of member countries, especially African (after achieving their political independence), that the great industrial nations have lost their decisive power in the General Assembly.

In the two important financial organizations that originated from the Bretton Woods Agreement of December 27, 1945 (the International Monetary Fund and the Bank of Reconstruction and Development, or World Bank, both with headquarters in Washington), the United States and the great European powers have maintained, since the foundation of these organizations, a disproportionately high participation in their boards and have exercised nearly absolute control in the orientation of their policies and administration, as if they were business companies in which shareholders had voting power on the board of directors in proportion to the number of shares held.

It is surprising that representatives of Third World countries have been attending, year after year, the annual meetings of the International Monetary Fund and the World Bank, practically as observers, passive listeners to the reports submitted by the executive directors and presidents of these organizations, who are officers appointed in practice by European countries (IMF) and the United States (World Bank), respectively.

The United States also exercises great power in the Inter-American Development Bank, an organism that was created in 1960 for providing financial and technical assistance to Latin American countries.

The United Nations Conference on Trade and Development (UNCTAD) was created specifically for studying such problems of Third World countries. Although Third World representatives at its periodic meetings have made vigorous statements related to the situation of dependency to which their countries are submitted, they have to date been unable to effectively implement their plans and recommendations, notwithstanding the fact that this organization represents more than 90 percent of world nations.

Likewise, developed nations have managed, through their great technological development and world political position, to exercise great influence over national and international public opinion. Through their majority interest in the cinema, television, international press agencies, great editorial firms, and a large part of the most widely circulated world press, they have been able to continually express their points of view to the underdeveloped world. The Soviet Union, for its part, has also taken an offensive in this field, but through its well-known traditional means of ideological penetration.

The controversy surrounding such organizations as the Central Intelligence Agency (CIA) of the United States is sufficient evidence of the immense power the intelligence agencies of the great powers have throughout the world and of the rather unorthodox methods they sometimes employ. For its part, the Russian KGB has for decades been a weapon of cold war, and in general, of Soviet expansion throughout the world. The activities of these and similar agencies have frequently been supported by the embassies and consulates of the Eastern and Western world powers.

The phenomenon of dependency, therefore, has truly deep roots that we are all under an obligation to analyze, serenely, shorn of all ideological and political influences. If industrial nations genuinely wish to understand the situation of underdeveloped countries and the causes of their dependency, it is necessary to assure that the latter's motivations are no longer distorted or presented as being contrary to the interests of developed nations.

When matters of nationalism, patriotism, sovereignty, and imperialism are discussed in underdeveloped countries, they are frequently interpreted by the public of industrial nations as manifestations of threats to their security. A healthy attitude would be to open the eyes of this public so they might realize that the prosperity and well being they enjoy is partly a consequence of the detriments being caused to underdeveloped countries. Furthermore, this inequitable situation from an economic viewpoint is a cause for the nationalistic manifestations so much feared. Responsible and practical leaders of industrial nations, if they genuinely desired to avoid serious world conflicts of a magnitude difficult to predict, should become conscious of the imperious need to revise and equilibrate their economic and trade relations with Third World countries.

Also, in underdeveloped countries the problem of dependency is sometimes inadequately focused. Some leaders fall under the influence of dogmas and ideas proceeding from developed nations, instead of focusing on the problem of dependency from the viewpoint of their own countries. What might be good and true for the developed world does not necessarily have to be so for the underdeveloped world. Also, and with certain frequency, many of these leaders in their struggle against economic penetration and domination by capitalist countries, commit the grave error of taking shelter under communist postulates which pretend to be supporting their cause but which, in practice, are only at the service of the imperialistic expansion of socialist countries.

As we shall see, Third World countries, with petroleum-exporting countries in the vanguard, identified by their condition of underdevelopment and dependency—and disregarding the political, cultural, religious, economic, and other differences that might exist among them—must become united in order to release the bonds of dependency to which they are subjected and to establish a new, more equilibrated, fair world economic system.

The Multinational Corporations as a Device for Economic Penetration

Within the world of private enterprise, multinational corporations are entities having very special dynamics and organizational structures, vested with objectives and projections that are distinctive in their formation and development. Because of the vastness of their organization and of the resources at their disposal, it is said that they represent the most efficient form of management ever structured up to the present.

"Multinational" does not mean that the companies are integrated by several countries, but rather that their activities are developed within and outside of their home country. Originally they were national companies located in their respective countries, but subsequently, and due to their expansion, they extended their operations to other countries through affiliates, subsidiaries, or other combinations. With the exception of special cases, their organization is essentially characterized by the centralization of company decisions at head office level, to such an extent that their subsidiaries are reduced to mere tools for executing those decisions.

It is not my purpose to make an exhaustive study of the nature and characteristics of these multinational corporations, since such an analysis would stray beyond the realm of this book. However, I do not wish to completely avoid the subject, because the idea prevailing today is that such corporations at present constitute a fundamental instrument in trade relations among developed nations and between these and underdeveloped countries. They are considered to be a highly important factor in the economic penetration of Third World countries by industrial nations.

Capitalist countries, headed by the United States, were the pioneers in creating this type of supranational corporation, for many reasons: their extraordinary economic expansion, always seeking new areas of influence; their wide range of technological development; their accumulation of vast capital available to be invested in other countries; their free-enterprise economy; and the circumstance that in this system—unlike the system of centralized economy, as in Russia—it is the private investor who takes the business initiative and is able to effect all kinds of transactions that are not reserved for the state.

The petroleum companies, as we have seen, by their very nature were among the first to originate and develop such multinational corporations. Activities developed in international spheres since the last century by John D. Rockefeller, Henri Deterding, and other American, Dutch, and English businessmen, enabled the formation and expansion of two of the most powerful and extensive conglomerates in the world: Standard Oil and Royal Dutch-Shell. Various others were developed later.

It is said that some of these corporations have more power than many countries. In support of this thesis I might mention the following facts: General Motors' sales exceed the value of Belgium's national production and closely follow that of Holland; furthermore, 4 million individuals have their destinies tied to that of General Motors. Exxon's sales exceed Switzerland's national production, Ford Motors' exceed production in Turkey, and those of General Electric exceed production in Sweden. According to financial publications, Gillette[4] is able to obtain world credits with far more ease than the government of Portugal. Mobil Oil, whose concessionaire in Venezuela holds sixth place as a producer of crude (slightly under 100,000 barrels per day) has an annual budget that is higher than the increased budget of the Venezuelan government. As stated by Sterling F. Green, "In financial terms, multinational companies are world powers in themselves."[5]

But of greater concern to students of this subject is the fact that these multinational corporations have become, both in underdeveloped countries and in their own industrialized nations, instruments of economic domination. Their activities go beyond the control of the countries wherein they operate, responding exclusively to the guidelines established by their own centralized administrations. I feel the subject merits some comment in relation to the manner in which these conglomerates operate today in underdeveloped countries, so that we can see how they exercise their economic power therein.

[4]This company, which sells in the millions of dollars per year bracket in Venezuela, thereby earning millions in profits, started activities in Venezuela with a capital contribution of less than $70,000. The remainder of its financial needs was obtained from local banks usually under parent-company guarantees.

[5]*El Universal,* Caracas, Jan. 21, 1973.

First, a process of vertical integration between the parent companies and their subsidiaries is established. This means that the subsidiaries' production, financing, and technology depend upon decisions adopted by head offices, and likewise, that the transfer of such elements to the receiving country is effected exclusively through these branches. To achieve penetration in underdeveloped countries, multinational companies have frequently adopted the following strategy: First, they export their finished products; they then establish sales organizations within the underdeveloped import countries; subsequently, they grant licenses to local manufacturers; and finally, they proceed to purchase or control many of the companies so established.

After completing this process of expansion and acquisition of local companies, the multinational corporations then provide from abroad a great part of top-level personnel in administrative and financial management, technology, design, marketing, and so on. Having accomplished these steps, they are then in a position to bypass the protectionist measures adopted by underdeveloped countries in an effort to confront their economic power.

For example, the policy of substituting local products for imports, which underdeveloped countries use to stimulate local industrialization, has often been blocked by multinational companies, because it is at their head office departments of marketing, personnel, investigation, planning, finance, etc., where the new products are developed, later to be assembled by their affiliates located in the underdeveloped countries. Furthermore, the new-product manufacturing systems, the machinery, equipment, innovation, and planning necessary for creating and expanding local markets, are also often centralized at the parent companies' headquarters. On the other hand, within the underdeveloped countries the multinational corporations have the necessary subsidiaries for routine marketing, assembly, and production.

From the foregoing it can be seen that the policy of substituting a truly local industry for imports, attempted by underdeveloped countries in order to diminish their dependency on developed nations, has sometimes become a tool serving the multinational corporations, which continue exercising their economic dominion over the underdeveloped world.

The multinational corporations also at times strengthen their position in emerging countries by taking advantage of measures adopted by those countries for their industrial development: market protection; exoneration from import duties covering equipment and raw materials; export premiums; local financing using public funds under subsidy terms or through private financial institutions; international technical assistance obtained by the host country, and so on.

Statistics prove that many of these corporations, after a small initial investment, have obtained loans and local financial resources in general, usually guaranteed by their own head offices, of great magnitude and under very advantageous terms. It is estimated that between 1963 and 1968 only 9 percent of the funds used by United States subsidiaries proceeded from their home country. In this manner, with financial investment from abroad continually decreasing in relation to the volume of investment and sales in the host countries, multinational companies have succeeded in accumulating high surplus accounts, based on their servicing of the highly protected local markets. As time has passed, this has placed strong pressures on the balance of payments of many underdeveloped countries, due to the payout of dividends and withdrawal of funds proceeding from asset depreciation and royalty collection by the multinational corporations.

Because of their economic power, multinational corporations are able to exert influence on host governments, which favor them in the distribution of certain infrastructure resources. On some occasions—such as in the ITT case in Chile, arising from the electoral triumph of Salvador Allende in 1970—multinational corporations directly intervene in the internal political affairs of the countries in which they operate, as they have done in their own countries, in order to gain special advantages. A wide range of literature exists on this subject, which is certainly interesting and revealing.

It is therefore logical that underdeveloped countries are constantly concerned about the control being exercised over their economic, social, and even political life by these corporations, and about the vast freedom these corporations enjoy in carrying out their commercial, industrial, financial, and other policies, to their own best advantage and interest.

In view of the preceding panorama, I feel the answers to the following questions are of extreme interest: What will the future of multinational corporations be? Will they continue growing and expanding? Will others appear? Will they undergo substantial changes? If so, will they always be promoted by the United States and the principal industrialized nations, or will certain rapidly growing semi-industrialized countries (e.g., Israel, Brazil, Mexico, South Korea) as well as the petroleum-exporting countries be added to the group? Or will these powerful international organizations lose force and tend to disappear?

Notwithstanding controls that may emerge at an international level as a consequence of increasing world interdependency (in turn the cause and effect of the multinational companies' expansion), there is a certain logic in assuming that these corporations will become stronger and increase in number. It is possible that regardless of the attacks and investigations to which they are often submitted in the United States, they will nevertheless be supported by the United States government in the future.

Petroleum-exporting countries, and the Third World in general, must become fully aware of the extraordinary power held by multinational companies; they are at present almost impregnable, due to the nearly unlimited freedom of action they have throughout the world and their financial and technological strength. However, true as it might be that up to the present these companies have been factors in the dependency of underdeveloped countries, the latter might now start using this same instrument intelligently to achieve their main objective of changing the present economic order and achieving an improved world equilibrium.

A genuine nationalism in petroleum-exporting countries, and in Third World countries in general, should help them attain more practical results, taking into account world orientation. Within this orientation, the tendency toward an increasingly closer world interrelationship appears irreversible, based principally on the economic and technological cooperation which in large part is being impelled by the multinational companies.

It might seem contradictory to believe that underdeveloped countries could be inclined to utilize, among other means, multinational companies, which are under such intense criticism insofar as their economic monopolizing activities are concerned. However, to reach such a stage of development, it might prove essential to resort to the same devices that were so successfully used by industrial nations—notably, multinational companies and the close international technical and economic cooperation they have promoted among industrialized nations.

To be able to live successfully within an even more integrated economic—and even legal—climate requires a gradual change in mentality, whether certain nationalistic leaders like it or not. It signifies the recognition and acceptance of the fact that such a change in attitude is already taking place in European Common Market countries, in the United States, in Japan, and even in the socialist bloc.

Petroleum-exporting countries, because of the abundant resources at their disposal, can and must open the door to allow economic interdependency to become the prevailing token in their relations with the group of industrialized nations. For that purpose, the creation, acquisition, and development of companies having a multinational structure, but originating in underdeveloped countries, might lead the way toward achieving this fundamental goal and also serve as a counterbalance to penetration and control by companies belonging to the great Western powers.

The Alarming Impoverishment of Several Countries in the World

One of the stark realities in the world at present is the poverty prevailing in many countries. The alarming deterioration in the economies of numerous Third World countries is rapidly leading them to a marked and progressive impoverishment. This reality should move the more prosperous nations to action and to assume the obligation of adopting emergency measures channeled toward exorcising the misery and death progressively enveloping the populations of these poor countries.

To fully understand this situation, it is insufficient to dryly repeat, as in just another journalistic report, that hunger and misery scourges a great part of mankind. Prosperous nations should, and must, bring this reality to light in the crudest possible terms, and formulate solutions.

The distinguished British scientist and novelist, C. P. Snow, at a conference held at Fulton, Missouri, toward the end of the 1960s, warned that the time was approaching in which "many millions of people in poor countries are going to starve to death before our eyes, and we will see them doing so on our television sets." The Paddock brothers in their *Famine in 1975,* published in the mid-1960s, depicted a similar panorama. In this excellent book, in view of the dramatically critical degree of the situation and using a drastic method reminiscent of that adopted with the wounded in times of war, they suggested classifying countries into three groups: those that could survive with due assistance; those that would perish in the long run whatever assistance might be provided; and those that would survive in any case. On the basis of this classification, the Paddock brothers recommended that, due to the limitation of available resources, assistance should be given only to the first group of countries.

In April 1974, Kurt Waldheim, Secretary General of the United Nations, tried to persuade prosperous nations to contribute $5 billion to help save the most desperate countries. Regrettably, by October of that same year he had only been able to obtain agreements amounting to $2.4 billion, of which $1.5 billion came from one country alone: Iran.

Experts of the Food and Agriculture Organization (FAO), a United Nations organism investigating world agricultural production, believe that thirty-two of world countries (i.e., more than one-fourth of the total) are so poor that they are struck by famine and on the brink of economic bankruptcy. Among these countries are Bangladesh, Southern Yemen, El Salvador, Ethiopia, Haiti, Honduras, India, Cambodia, Laos, Nigeria, Pakistan, Upper Volta, and Yemen.

As we know, in many of these countries there are great disparities in geographical regions, which means that substantial sectors of their populations are living in conditions by far more tragic than others and are, in fact, dying from starvation. A dramatic example of this poverty and death is Bangladesh, a country of the Hindustan Peninsula, whose population is currently suffering the most dreadful crisis of famine in thirty years.

As predicted by C. P. Snow, world press and television reports have dramatically shown the setup of camps in Bengal for famished people, guarded by local police in order to ensure their staying within. These ghettos house a multitude of starving women and children whom the government of Bangladesh does not wish to have straying about the capital, because their presence provokes social tensions.

A tremendous flow of humanity arrives daily at Dacca by ship, by train, and on foot, in search of the food that they were unable to find in their home regions. At passenger terminals and other zones of the city, people lie on the ground and along the sidewalks, exhausted from hunger; they are then picked up by the authorities and taken to the so-called purée kitchens, created by the government for providing emergency rations to more than 5 million peasants (out of the 77 million the country has), who day by day face death through starvation. There are 4,000 to 5,000 kitchens of this type functioning throughout the country, but they are unable to meet the crisis. According to news arriving from Dacca and from reports submitted by voluntary aid organizations and by some foreign diplomats, it has been conservatively estimated that in 1974 not less than 1 million persons died of starvation or related illnesses in Bangladesh. The situation is so drastic that many have resorted to suicide after devouring their last "meal."

But famine in the Hindustan Peninsula is merely an example of what is happening in many Third World countries. Robert S. MacNamara, President of the World Bank, at annual meetings of that institution and of the International Monetary Fund, and also in his book *One Hundred Countries, Two Million People,* has pointed out that from the 2 billion inhabitants of developing countries, 40 percent, i.e., approximately 800 million persons, barely survive in a marginal existence, under dreadful subhuman conditions of sickness, malnutrition, and misery. The degree of poverty to which men, women, and children are subjected is so intense, says MacNamara, that they are ensnared in a situation in which it is impossible to develop any productive potential whatsoever.

Many are the causes of this impoverishment. In the first place the decrease in world harvests in recent times is mentioned. The limited agricultural yields of 1971 and 1972 put an end to the euphoria then prevailing over the so-called Green Revolution, consisting of the so-called miracle seeds which doubled harvests in some regions of Latin America. And, in 1973 and 1974, poor harvests were registered in a great number of countries throughout the world.

The economic boom occurring in recent years in industrial nations has also increased demand for food and consequently its shortage. The United States, which has traditionally been the world's principal granary, has witnessed a dangerous decrease in its food stocks since 1970 principally as a consequence of sales to Russia of approximately 20 million metric tons of wheat and 10 million of coarse grains. Many believe these sales were an unconscientious act, especially considering that the United States was unable, in 1973, to rebuild its stocks or to incorporate idle farmland into cultivation. By February 1974, the food shortage had caused a substantial rise in the United States export price of wheat, which, compared to June 1972 levels, had increased fourfold. In Bangladesh, prices of grain doubled between January and August 1974, and in Thailand the price of rice has doubled since August 1973.

FAO experts estimate that, taking the 1970 food per capita in developing countries as a basis, the deficit in the production of grains in 1980 will be on the order of 50 million tons. As a result, the world faces its food shortage in history. The increase in food prices is far above the economic means of many Third World countries; many poor countries are spending between 70 and 80 percent of their income on food imports. It goes without saying that such substantial price increases represent less nourishment, increased malnutrition, and consequently, an increase in the number of deaths through starvation.

To the basic causes that I have just mentioned, which have accentuated the impoverishment of several countries, I must add two others: the inequitable balance of world trade and the explosive population growth in Third World countries.

The world awakens each day with more than two hundred thousand new inhabitants demanding their share of food. More than 70 percent of these infants are born in underdeveloped countries, which, as is known, have the lowest resources available to meet their nutritional needs. The population growth in these countries has been such that in the last ten years they have spent more than half their income on food. It is important to have a clear knowledge of what explosive population growth will mean to the world in forthcoming years, and especially, of the fact that this growth is mainly occurring in poor countries, which double their population every 20 years. This means that the essential needs of those countries are also doubled, because population growth automatically accelerates consumption, atmospheric pollution, exhaustion of limited natural resources, and so on. It also represents a challenge to the governments of these countries, which will have to provide their inhabitants with adequate supplies of water, food, housing, social assistance, educational facilities, sources of employment, i.e., all the essentials for a decent standard of living. If these foreseeable effects are not counteracted, this impoverishment might lead to extremely dangerous political and social tensions in the world. We are already able to see that this serious problem affecting the Third World is more liable to become critical now than it was thirty years ago, because more than half of the world's population today is under fifteen years of age. This young population, located principally in the poor countries, will soon seek—desperately in some cases—the ways and means for achieving better standards of living.

The obvious facts set forth in the preceding paragraphs demonstrate that a great number of countries belonging to the Third World are markedly and progressively becoming further impoverished and will be unable to overcome this situation without assistance from countries that at present have the means of production and sufficient financial resources to come to their aid.

It therefore seems surprising that, as mentioned in my introduction to this chapter, where I quoted OECD information sources, the small group of rich nations, notwithstanding a 60 percent increase in their net income during the ten years from 1963 to 1973, have decreased their foreign aid by 11 percent in the same period.

These nations should discontinue their avid consumption, in such a great proportion, of the natural resources proceeding from poor countries and acquired at minimal subsidy prices. For their part, petroleum-exporting countries—although obviously not having been the generators of the present world inflation or of the alarming impoverishment of several Third World countries—must become conscious of the undeniable fact that the recent substantial increase in oil prices has partly contributed to aggravating the misery prevailing in the poor countries on earth. For the less developed nonpetroleum countries, the cost of their present volume of petroleum imports has increased by $10 billion, according to expert estimates. This new factor will further complicate their difficulties in financing the high deficits in their balance of payments.

Because of this situation, petroleum-exporting countries must bear in mind the need to cooperate, in many ways, with other underdeveloped countries, especially the most impoverished, as an act of solidarity and humanitarian duty. And I reiterate that these petroleum-exporting countries need support by the Third World countries in their common struggle to override the ties of economic dependency to which they have been subjected by a small group of industrial nations, a struggle that has successfully commenced through the increase in petroleum prices, a tendency that might also be applied to other raw materials.

The Change in the Equilibrium of Forces in the Petroleum World

To outline the fundamental goals that Venezuela and other petroleum-exporting countries might pursue in their oil policy and international strategy, I believe it is first necessary to analyze the most important aspects characterizing the world economic situation in the 1970s.

Dependency and interdependency, development and underdevelopment, impoverishment of Third World countries and economic recession in rich nations—these are the factors that currently occupy the minds of those engaged in the analysis of world problems.

To these must be added the change in the equilibrium of forces that has taken place in recent years in the oil world between petroleum-exporting countries and the oil companies, and between these countries and the great developed nations, consumers of hydrocarbons. The balance of power, formerly inclined toward the oil companies, is now oriented toward the petroleum-exporting countries, which, upon assuming the power of decision in world petroleum policy, especially in price setting, and upon obtaining higher income from the sale of hydrocarbons, are generating a change in the equilibrium of forces within the petroleum world.

I referred in detail to this change in the equilibrium of forces in Chapter 9, under the heading "The Era of Nationalism," because I believe that a full comprehension of this new reality is of interest to Venezuela in particular, and to petroleum-exporting countries in general. I wish only to add that this new situation should be the foundation for the objectives and strategies forming the structure of the oil countries' foreign and petroleum policies.

Economic Interdependency in the Developed World: The Socialist and Capitalist Blocs

Parallel to the nationalistic effervescence developing intensely in most Third World countries, there is another phenomenon of major importance gaining force: the economic interdependency among developed nations. This process is becoming clearly defined in the Western industrial nations. It is thus that the tendencies and economic variances in each one of these nations are being strongly felt by the others, and the prosperity or economic depression in one given country is easily exported to the others.

Very schematically, and to illustrate this phenomenon, one could say that a decrease in Japan's economic activities would cause a decrease in its imports of raw materials from Canada. This situation, in turn, could weaken Canada's incentives for exploiting new mines, thereby forcing it to cancel its orders for mining equipment from Germany and France, which, in order to preserve their balance of payments, would decrease their imports of computers from the United States.

From the preceding example it can easily be seen that the chain is unending. Furthermore, it is submitted to a multiplying factor in the Western industrial powers because of their increasingly stronger economic ties since World War II. Today these ties tend to be further strengthened as a consequence of the relative shortage of petroleum supplies, the fear of future embargos on such exports, and the substantial increase in petroleum prices, which is draining the monetary reserves of industrial nations and affecting their balance of payments.

This situation of economic interdependency also affects Third World countries because of its world projections. However, I prefer to consider interdependency as a phenomenon of the developed world, so as to avoid confusion in my analysis of the situation prevailing in underdeveloped countries, which, rather, are being submitted to the state of dependency on industrial nations described earlier. Interdependency implies the existence of a basic freedom among the parties involved, mutual cooperation and equal negotiating capacities, as in industrial nations; whereas dependency implies a dominated party on the one hand and a dominant party on the other, as between underdeveloped countries and developed nations.

It is interesting to see that economic interdependency tends to spread among all industrial nations in the world, causing political differences to lose a certain degree of importance, because economic combinations seem to override the ideological barriers existing among them. Only a few years ago the world was impressed by the news of an agreement between the Soviet Union and Fiat for the development of an important automobile plant. Subsequently, other Western industrial companies participated in the construction of chemical plants and other modern installations in the Soviet Union. The Soviet Union has also opened the door for closer collaboration in the oil field, as proved by the scientific-technical agreement of cooperation signed between the Soviet government and Occidental Petroleum Corporation (OXY) in 1972.

It seems quite obvious that the Soviet Union is confronting certain technological problems in the development of its petroliferous potential, principally in Siberia, to the east of the Ural Mountains, as well as in other fields of endeavor. Apparent lack of qualified personnel and capital could be the reasons that the Soviet Union accepted participation of Japan within its country and, more recently, the aforementioned Occidental Petroleum.

Besides OXY, other groups of United States companies—such as Gulf, El Paso, Tenneco, Texas Eastern, Bechtel and Brown, and Root—seem willing to participate in the exploitation of hydrocarbon resources in the Soviet Union, and to receive for their services payment in kind in the form of crude or liquefied natural gas, thus revealing the avidity with which United States petroleum groups are seeking sources of supply due to the increasingly acute crisis of energy supplies in the United States. This economic cooperation between both blocs became feasible—and was even propitiated—when the policy of coexistence was promoted by President Richard Nixon.

To further illustrate this rapport between the Eastern and Western industrialized blocs, I cite the following trade and industrial combinations between companies of the Communist bloc and the Western bloc:

The Swedish chain of furniture stores, IKEA, supplies its partners in Poland with machinery and design for partial manufacture of furniture; these semifinished products are then shipped to Sweden, where IKEA completes them.

The Italian state company, Meccania Saetana, and the Bulgarian company, Machenoexport, have created a company in Switzerland (ZOCCA) on a fifty-fifty ownership basis, for world marketing of grinding machines. Zocca receives semifinished machines from Bulgaria, completes their manufacture, and sends the finished product to the marketing company. Zocca provides its services for selling the machines in the Western world; Machenoexport intends to do the same in the Socialist bloc of countries and will be in charge of selling all Zocca products therein.

The United States Simmons Machine Tool Corp. has drawn up an agreement in which the renowned Czechoslovakian Skoda company will produce a line of Simmons' heavy products, under the trademark Simmons-Skoda. Simmons estimates that deliveries will be cut down from fourteen months to approximately one, in view of the Czechoslovakians' willingness to maintain a stock of machines, a practice which generally is not followed in the West for this type of product.

Krupp of Essen, Germany, and the Csepel Tool Factory of Hungary have jointly developed small digital-controlled lathes; these products will be exported to Germany and to other markets.

Pechiney St. Gobain (PSG) of France and the Progress Invest of Belgrade will jointly manufacture and deliver whole plants ("turnkey jobs") of chemical fluoride fertilizers and organic and plastic chemicals. The plants will be built in Yugoslavia, France, and other countries.

The foregoing clearly illustrates the tendency toward economic and technological collaboration among nations (especially those that are most developed), independent of their fundamental ideological differences.[6] Thus, agreements, ventures, and combinations of all types, especially in the industrial, technical, and commercial fields, are becoming intensified within the Western developed world and expanding to the Socialist bloc. They are the consequence of conversations between companies of both blocs, where ideological differences, propaganda, and political and military threats are left aside. The way is clear for concrete agreements covering the joint development of economic, technological, commercial, and financial projects, and also others of a humanitarian and artistic

[6]This tendency has not, however, prevented the Soviet Union from instigating and financing certain sabotage maneuvers within Western countries' economies, with the intention of weakening their competitive strength. An example is the numerous strikes and labor conflicts that have besieged Italy, causing an increase of around 100 percent in salaries in 1972, thereby weakening the rivalry of Italian products against those of the Socialist bloc.

nature, for the purpose of achieving common objectives conscientiously and deliberately stripped of any ideological content whatsoever.

As a consequence of the drainage of certain products and basic raw materials (in some cases critical), many industrial nations are becoming convinced of the need to apply, on a more rigorous basis, the theory of dividing work internationally. According to this thesis, each and every product would not be developed and manufactured in each country individually, but within a specific group of industrial nations, or within determined spheres of political and ideological influence.

In support of this thesis we have the extremely alarming phenomenon of pollution generated by intense industrial activity in the most advanced nations. Over a relatively short period of time, such pollution will possibly force them to divert part of their industrial expansion toward other nations. It would not be surprising, therefore, if in forthcoming years a transfer of industrial centers is made to less developed countries, where adequate ecological conditions prevail and sufficient raw materials are available. In some European countries of great industrial concentration, such as West Germany, scrap proceeding from its industries is virtually transforming some of the rivers crossing the industrial zones into sewers. Furthermore, as a consequence of Western Germany's superindustrialization, there is a shortage of manpower. Laborers are being brought in from other, less advanced countries, such as Spain and Portugal, and from the Third World, thereby generating problems of a social nature. In Germany, temporary immigration of labor exceeds 2 million individuals.

In addition, traditional export sales organizations and policies are no longer adequate for maintaining universal acceptance of finished and semifinished products, and even raw materials. It is becoming necessary to establish selling, manufacturing, and development units at the consumer centers. Therefore science and technology, as well as administrative talent and even labor, are being increasingly mobilized on a worldwide basis.

As well expressed by Howard Perlmutter, professor of social architecture at the University of Pennsylvania:

> While the profound implications of a nuclear disaster gradually invade our minds, the impassioned adherence to ideologies is being revised both in the East and in the West. The results of the different outlooks for structuring a world economic system are becoming more apparent and human interdependency is becoming more concretely felt in a process of industrialization that directors of multinational companies must administrate.
>
> There exists a process between the East and the West that should legitimize what we call trans-ideological zones, by means of trans-ideological dialogue, in order to culminate in a greater abundance of systems, already existing, and covering licenses, subcontracts and co-production. . . .

The fact that the world economic system must be a joint system, and that interests truly in common exist, is becoming increasingly accepted.

There exists no other way for maintaining unity on earth, other than by admitting that the common ideology uniting humanity in its present struggle for survival and for improved living conditions is not the special prerogative of either the East or West.[7]

In 1968, President Richard Nixon asserted that the era of confrontation with the Communist world had ended and that it was being replaced by an era of negotiations. "Whoever the United States President may be in the next four or eight years," he declared, "he should proceed under the assumption that negotiations must take place with the Soviet world and, eventually, with the next superpower: Communist China. This is a change that has occurred and therefore his policy must change accordingly."[8] Events have demonstrated the accuracy of this prediction.

During a trip to Washington and to Western Europe (France), Soviet leader Leonid Brezhnev apparently sanctioned the new type of relationship that is becoming established between the Soviet Union, on the one hand, and the United States and Western countries in general, on the other, i.e., between the East and West. It seems that while politics still divides them, economics tends to unite them. We may therefore conclude that substantial common economic interests are becoming intertwined. Financiers of Wall Street and other Western economic centers, on the one hand, and the technical leaders of Moscow, on the other, are discovering that their respective systems are complementary in many aspects.

Since it is not convenient at present for either of the two major world powers to definitely predominate over the other, they are both visualizing reciprocal interaction within a cooperative effort, such as the Siberian pipelines and industrial technology, as ideologically neutral symbols of a new international pact. This interaction does not imply that Russian-American relations, and those of the East and West in general, will not substantially remain in an atmosphere of considerable mistrust, as such attitudes are comprehensible between two counteropposed systems. Incompatibility of principles still prevails between the two ideologies and the two systems; within the work plan, the most deeply ingrained competitive tendencies are evident at a planetary level. However, Russia, fully

[7]Howard Perlmutter, "Emerging East-West Ventures: The Trans-ideological Enterprises," *Columbia Journal of World Business*, vol. 4, no. 5, September–October 1969.

[8]Statement by Richard Nixon, reported by the *Indianapolis Sunday Star*, Aug. 18, 1968; quoted by Gary Allen in *Richard Nixon: The Man behind the Mask*, Western Islands, Belmont, Mass., 1971, p. 2.

conscious today of its technological lag and of the heavy bureaucracy gravitating toward many of its production sectors, not only needs to absorb and to apply new industrial techniques, but also needs the elements proper that integrate the modern administrative processes applied in the West. These are the factors that are pushing Soviet leaders over the barriers of ideological discipline, to normalize their relations with the United States, Europe, and Japan, and to endeavor to overcome the former "equilibrium of terror," at the cost of lesser sacrifices.

In this new order of affairs, the main counterpart for Western powers (which are warming to the decreased risk of world war and to the increased suffocation of armed conflicts that have repeatedly struck humanity since World War II) lies in the imbalance represented by the existing critical exhaustion of their primary resources (especially in the United States), compared to the tremendous economic potential of the Soviet Union, particularly in raw materials. The Soviet Union controls more than one-seventh of the earth's resources.

Thus the interdependency between both blocs is only relative. It is fundamentally circumscribed to certain economic ventures, such as those described previously, the volume of which is as yet relatively small, and to trade agreements that are sometimes very slow in becoming perfected and that are still tied to political considerations. In fact, trade exchange between the United States and the Soviet Union was in the ratio of 4½ to 1 in 1963, when United States exports reached $1.1 billion and Russian exports were only $244 million. This ratio dropped to less than 2 to 1 in the first eight months of 1974, however, due especially to the completion of United States grain shipments to Russia.

Regarding trade agreements between these two countries, in 1972 the Nixon administration promised the Soviet Union nondiscriminatory traiffs and large-scale trade credits under advantageous conditions. However, this promise was not fulfilled because of opposition by the United States Congress. The long delay by Congress in pronouncing its decision particularly irked the Soviet Union. Finally, approval was granted in exchange for what appeared to be the Russian government's agreement to soften its emigration policy and allow the exodus of at least 60,000 Jews and other racial minorities per year, according to an estimate made by Democratic Senator Henry M. Jackson, who was one of the principal participants in the supposed agreement and in shaping the congressional law. The agreement also included the Soviet Union's willingness to constructively consider any United States claim resulting from lack of Russian compliance—in the opinion of the United States—in measures tending to soften the Soviet emigration policy.

If true, this agreement implies interference by the United States government in the Soviet Union's domestic affairs, a policy difficult to understand. The agreement was, in fact, originally reported by United States Secretary of State Henry Kissinger. (At the last moment, the day before the law was approved, the Soviet Union astutely proclaimed that no real assurances had been given after all.)

This apparent interference in the Soviet Union's domestic affairs, which was criticized as being inconvenient to the practical application of the 1972 Soviet-North American Trade Agreement, reached a crisis in January 1975. At a press conference in Washington on January 15, Secretary Kissinger stated that the Soviet Union had communicated to the United States that it would not apply the 1972 Trade Agreement because the U.S. Congress had imposed the requirement that Moscow should allow free emigration of Jews. It is said that this rejection is based on the fact that the Soviet Union considers the new United States foreign trade law to be contradictory to the objectives of the 1972 Trade Agreement, which included the elimination of all discriminatory trade restrictions, as well as contradictory to the principle of nonintervention in domestic affairs.

Although both of these arguments could be quite valid from the Russian viewpoint, I am inclined to believe that the 1972 Trade Agreement was rejected principally because of the intervention by the United States government in the Soviet Union's domestic affairs. The latter's attitude might also have been influenced by the wave of negative criticisms and reactions in Third World countries to the aforementioned law.

Although incipient economic interdependency between Eastern and Western industrial nations has in recent years become a reality, the events I have just described make it now more difficult than before to assume that trade and economic relations between both blocs will become closer, at least for quite some time. Furthermore, although exchange between the East and the West has taken place over and above ideological differences, the anchorage and expansion of these relations will depend considerably on the political interests of the Soviet Union and the United States in international fields, especially in relation to world petroleum (principally represented by the Middle East) and to Third World countries, whose move toward economic freedom seems already factual.

As we shall see later, the Soviet Union may be obliged to assume a more defined position in relation to its political and eventual military support of Middle East petroleum-exporting countries confronting pressures or belligerent attitudes that might be exerted upon them by Western powers in general, for reasons of petroleum prices, among others. Inversely, this could also be the position assumed by the United States, if the Soviet Union were to attempt to increase its economic, political, and military influence in the Persian Gulf countries.

Dependency on Raw Materials by the Western Industrialized World

As a consequence of their high indexes of economic growth and standard of living, industrial nations in the Western world are exhausting their own natural resources and becoming increasingly dependent on supplies of raw materials proceeding from underdeveloped countries. Without these raw materials today, they cannot keep their industrial machinery in motion; they cannot expand their economies or meet the levels of consumption to which their citizens are accustomed.

This fact has dramatically emerged in relation to petroleum. The substantial increase in petroleum prices imposed by OPEC in 1973, and the embargo by Arab states on exports to the United States and to European nations at the end of that year, was a concrete symptom of this phenomenon, polarizing world reactions.

Prior to the embargo, a report of the Club of Rome[9] moved world opinion and caused controversy. Although the report includes debatable projections, it presents an excellent illustration of what is happening in the world in this field. It is pointed out that:

Where food is concerned, there has existed a great excess of potentially arable lands throughout history. However, in the next thirty years, when world population conceivably doubles, it seems inevitable that a tremendous food shortage will occur. A considerable amount of what in principle is arable land—but very little, in relative terms—belongs to underdeveloped countries, which at present have precarious levels of nutrition.

As to minerals and vital energy resources for present world industrial development (aluminum, chrome, copper, carbon, cobalt, gold, iron, tin, manganese, mercury, natural gas, nickel, petroleum, platinum, and zinc), and taking into account current rates of increase and multiplying known reserves by five, we see that, should humanity continue consuming such products at present rates, most of them will become exhausted within the next sixty years.

From statistics given in that important work, we can also see that the major consumers are the great developed nations. The main producers (except Russia and the United States, the latter in general consuming more than it produces) are the underdeveloped countries.

Therefore, taking into account the volume of reserves of these basic products, and the additional future reserves that might reasonably be estimated, this situation appears in all respects unbearable for the world. It is especially unbearable for developed nations, the majority of which

[9]See *The Limits to Growth*, 7th ed., Universe Books, New York, 1972.

(such as Japan, European countries, and now the United States) depend upon their imports of Third World raw materials in a excessively high proportion. The crisis is even more serious when we reflect on the fact that this extraordinary shortage of basic raw materials, both renewable and nonrenewable, mainly possessed by underdeveloped countries, has already occurred and is even tending to become aggravated as a consequence of increasing consumption by the small number of developed nations.

We can find a suitable example of this phenomenon in the case of Japan, which is the second most industrialized nation in the world. For further illustration, I shall quote the authoritative words of Saburo Okita, an outstanding expert in economic affairs and a prominent Japanese public personage, from his report: "Natural Resources, Dependency, and Japanese Foreign Policy." In this report, he demonstrates through very precise figures Japan's dependency on imports of nearly all the natural resources fundamental for maintaining its economy and even for feeding its population. He states the following: "The current oil crisis has once again demonstrated to Japan her high dependency upon supplies of natural resources from abroad. When these supplies of foreign resources are obtained smoothly, the Japanese economy progresses favorably; however, once imports are interrupted, the impact on the Japanese economy is immediate and severe."

Furthermore: "Now the world economy must face changes such as the strengthened bargaining power of the resource-exporting countries, the intensification of so-called 'resource nationalism,' and limits in world resource supplies." He then ends by asserting the following (which is paradoxical if we consider that only a few years ago the great futurists— absolutely certain of Japan's economic potential—were sure that Japan would reach its postindustrial stage within a few years):

> With respect to the international aspects, what are the proper diplomatic policies for Japan to pursue? In Japanese *kendo* (fencing) terminology there is a posture called *jappo-yabure* implying "defenseless on all sides." As described earlier in this article, Japan's dependence on imports of raw materials, energy and food is so complete that policies attempting self-sufficiency in any of the key items appear unrealistic. Diversifying sources of supply, economizing on the use of raw materials and energy, stepping up efforts for increased production from indigenous resources, and building up emergency stocks of energy and food—all these are feasible and should be pursued with seriousness. But the basic character of the heavy dependence for key items on overseas resources will not change.
> One of the policy directions stemming from this fundamental condition is the pursuit of a diplomatic policy of being friendly with everybody, or at least

not making serious enemies anywhere. In order to make *jappo-yabure* an effective diplomatic policy, Japan must avoid becoming a danger to any other country in the world. If Japan started to build up military strength then at least some other countries would interpret this as a dangerous sign and in turn fortify themselves militarily vis-a-vis Japan. This might touch off repercussions which would be inconsistent with the basic vulnerability of Japan's resource base.[10]

In relative terms, the situation of economic vulnerability of Western European industrial nations, although less critical, is fundamentally the same. The United States is also vulnerable today, although in lesser degree. The world has become accustomed to thinking about the remarkable natural resources of the United States, which at one time carried it to self-sufficiency and to becoming the major export country in the world. However, as a consequence of the increasing and uninterrupted use of those resources, necessary for the United States economy and population, this country—the most industrialized on earth—has become dependent on petroleum and on a series of other basic resources. Furthermore, it is well known that even its traditional position as a major exporter of surplus grains and other foodstuffs has become weakened in recent times.

The potentially least dependent among the world's industrialized nations is the Soviet Union.

This dependency by developed nations on petroleum and other raw materials owned by Third World countries should be taken very much into account by OPEC countries when formulating a policy and coherent strategy in respect to the orientation, volume, and pricing of their exports, and also in relation to the development of their petroleum industries and to their position as obligatory leaders of Third World countries. Bearing in mind Russia's relatively more comfortable situation insofar as energy and natural resources are concerned,[11] it seems doubtful that this power—notwithstanding its present policy of détente with the United States—will be inclined to decisively support Western industrial nations in confronting petroleum-exporting countries and Third World countries in general, especially in the extreme case of belligerent action by the former against the latter.

This consideration is today crucial for petroleum-exporting countries, especially those in the Middle East, because of the possibility of military intervention. An illustration is the statements made by U.S. Secretary of

[10]*Foreign Affairs*, July 1974, p. 714.

[11]At the beginning of 1975, the Soviet Union announced that it had exceeded the United States' volume of petroleum production, having therefore become the major oil producer in the world.

State Henry Kissinger in *Business Week* (an extract thereof having been published in advance on January 1, 1975, by the State Department), in which he stated that the United States would consider the use of military force for solving the Middle East petroleum problem only in "the most drastic of emergencies," which he defined as "the strangulation of the industrial world." Two days later, January 3, the Presidential press secretary asserted that Secretary Kissinger "reflected the President's position" in stating that the United States would consider the use of military force only "in the gravest emergency."

Joint decision and action by OPEC countries would be required, since the United States and the other Western powers would be, according to the above declarations, the judges of when a status of strangulation existed in their economies, and they would be the ones to sound the bugle for military intervention. It is interesting to note a statement in Teheran's newspaper *Ettelaat* of January 1975: "Military intervention by one superpower would create the intervention of another, and the result would be nothing short of world tragedy."

These statements by United States official spokesmen reverberated throughout the world, awakening alarm over events that might lead to a third world war. For instance, Pope Paul VI, in his January 1975 annual message to accredited diplomats at the Vatican, acknowledging New Year's greetings, warned that the prevailing petroleum dispute could provoke a war. This declaration has been interpreted by observers as a response to Henry Kissinger's statements regarding the possible use of United States military force for preserving petroleum supplies. In his message, the Pope appealed to all world leaders to adopt courageous measures and preserve wise diplomacy in confronting the new and more threatening complications arising from the so-called War of Energy Sources.

Within this order of ideas, underdeveloped countries should bear in mind that in spite of the surging economic interdependency between the East and West rival blocs, any decline of Western industrial nations due to the deterioration of their economies (partly a consequence of the increased cost of petroleum and other natural resources imported from Third World countries) helps fortify the position of the Soviet Union and its satellite countries in the world equilibrium of forces.

Therefore, if petroleum-exporting countries are able to adopt a wise position facing the Soviet Union—although naturally maintaining their full status of independency—it appears unlikely that Western nations will obtain Russian support should they decide in the future to take any extreme direct or indirect action of a belligerent nature against those countries.

On November 1, 1974, the world press published a significant denunciation by *Pravda,* to the effect that the Western capitalistic world had launched "psychological warfare" against petroleum countries to attempt to force them to lower petroleum prices, "following the lines of the old cold war policy which poisons the atmosphere of international détente and retards world progress." It furthermore inquired whether or not "it would be possible to reach a compromise if the friendly relations between the Arab countries, the Soviet Union and other Socialist countries became undermined."

Yasir Arafat, the guerrilla leader of the Palestinian Liberation Organization, during his visit to New York in November 1974 to attend the United Nations General Assembly debate concerning Palestine, in an interview by *Time* magazine, warned that the Soviet Union would intervene if the United States militarily attacked Middle East oil fields, adding that "it would be an extremely deceptive calculation by the United States to think that other superpowers would not intervene."

Within this potentially critical reality, there are two favorable aspects. The first, advantageous in all respects, consists in the fact that oil countries, and Third World countries in general, today possess a powerful tool in their petroleum and raw materials for introducing a new world economic order. The second, of a relative character, consists in the fact that great industrial Western nations—the principal consumers of petroleum and other raw materials—have democratic political administrations which, in contrast to the Soviet Union's political administration, are sensitive to public opinion and to human rights. This aspect is a factor that could contribute to preventing a military intervention on the part of these nations.

Economic Recession in the Developed World: An Unforeseen Phenomenon

By owning great natural resources in some cases and through toil, discipline, tenacity, technical preparation, education, and many other qualities of their people in other cases, and as a result of favorable trade policies, a small group of Western countries emerged as great powers.

After World War II many of these nations, prostrated by the aftermath of war (especially Germany, Italy, France, and Japan), were able to recuperate—and even exceed—their former industrial status. Becoming great powers, they dominated world economy jointly with the United States. The flourishing economic growth of these powers allowed them to develop over a short time, especially in the United States, a vast technology that enabled them to consolidate their power at all levels.

On the other hand, prosperity created a consumer propensity in their populations, propitiating a standard of living (and waste of assets) never before known in the history of mankind. Many are the books and studies that have emerged concerning consumer societies, focusing on these aspects and their consequences, especially that of atmospheric pollution and the exhaustion of natural resources on a worldwide scale.

I must digress a little at this point to comment upon how scientists, politicians, and leaders sometimes incur mistakes in the euphoria of their success, prosperity, and power. Futurologists emerged in the 1960s, on what may be called an organized scale. They were unlike prophets in the past, who resorted strictly to their imagination and described, as in science fiction novels, what the world of the future would be, voicing prophecies unworthy of serious discussion because they were based on projections having little scientific foundation. The prophets of the 1970s were intellectuals of the United States and other advanced countries, closely allied to science and technology. By converting such sciences into a profession of faith, almost a religion, they used them to outline—with apparently irrefutable logic (at least for students of the subject, who associated them with the high priests of ancient Egyptian civilization)— the future of the great industrial powers, or the future of a world in which these superindustrial powers would prevail.

It was thus that these futurologists used their science of prospection and promoted, with financial support from governments and great corporations of industrial nations, several nonprofit organizations and study groups for the development of this science, which arrogantly professed to project, on what appeared to be a solid scientific and technological basis, the long-term future of the world. On the basis of highly refined techniques and methodology, they also felt it would be possible to control to a certain degree future events in order to achieve in practice fulfillment of the predicted end result.

Daniel Bell, chairman of the Commission on the Year 2000, of the American Academy of Arts and Sciences, asserted the following in 1967:

> If there is a decisive difference between the future studies that are now underway and those of the past, it consists in a growing sophistication about methodology and an effort to define the boundaries—intersections and interactions—of social systems that come into contact with each other.
>
> Every society today is consciously committed to economic growth, to raising the standard of living of its people, and therefore to the planning, direction, and control of social change. What makes the present studies, therefore, so completely different from those of the past is that they are oriented to specific social policy purposes; and along with this new dimension, they are fashioned self-consciously, by a new methodology that gives the promise of providing a more reliable foundation for realistic alternatives and choices, if not for exact prediction.

We can find a prototype of this ultramodern prophet in Herman Kahn, who together with Anthony J. Wiener wrote the famous report "The Year 2000." It leads to meditation to see how these foremost futurologists were occupied almost exclusively in prospections of a handful of industrialized nations. Their method of analyzing societies and their future embraced these nations alone, and many people acquired absolute faith in the validity of these projections, even though they referred to distant periods of twenty, thirty, and more years hence.

Out of this new school of thought, surrounded by an aura of technology, a newly conceived terminology emerged for describing the world of the future; e.g., "postindustrial societies," which refers to the level of development that would be reached at the turn of the century by the societies of a handful of privileged nations, such as the United States and Japan. In this projection one of the greatest problems that would be confronted was how to fill the idle time that would be available to the members of these societies.

It seems, however, that these futurologists overlooked or underestimated the fact that these nations, however advanced they might be and however great their technological possibilities, exist on a limited planet in which more than two-thirds of the population live in more than 100 underdeveloped countries; that in these 100 countries, approximately 2 billion people exist on levels close to subsistent and are affected, in greater or lesser degree, by malnutrition and famine; that in these 100 countries, the major part of the natural resources necessary for sustaining nearly all world population is found; and that the industrial machinery of the small group of rich nations greatly depends on these natural resources.

This fact was not taken sufficiently into account by these futurologists, who emerged a few years ago and whose investigations, books, lectures, and reports amazed the world, carrying them to the pinnacle of the most refined circles of world intellectuality. They furthermore seem to have underestimated the true importance of such evident factors as inflation in great industrial powers and the progressive deterioration of the international monetary system, which, since the Bretton Woods Agreement, these powers have been dominating.

Nor did the futurologists perceive that a group of underdeveloped petroleum-exporting countries, united in OPEC (which by then, 1960, already existed) would become more consolidated and gather sufficient force to impose, over a short term, much higher prices for their precious energy resource, thereby contributing to curbing the high index of economic growth that the rich nations had been experiencing.

Finally, they did not sufficiently consider that the other Third World countries, producers and exporters of all manner of products greatly important to industrialized nations, would possibly not take long in following the petroleum exporters' policy.

At the turn of the 1970s, the world was again impressed, not by the "postindustrial societies" predicted by these prophets with the assurance they felt through their vast technological knowledge, but rather by a crude and undeniable reality: the grave economic problems being faced by the industrial powers. These problems were a consequence, among other factors, of their dependency on petroleum and other raw materials no longer available to them, as in the past, at depressed prices and under the traditional trading terms that had so fully favored them throughout so many years.

The steep increase in petroleum prices and the embargo on exports at the end of 1973 were an alert to industrialized nations for them to realize that their development and exaggerated increase in consumption had certain limits and could no longer continue at the cost, in certain part, of poor countries.

This alert also applies to Third World countries. Although having justification, and in part, power on their side, these countries must realize that their attempt to emerge from the situation of exploitation to which they have been submitted does not constitute a utopia. They are under the obligation of establishing—because it will not come about freely—a more equal economic system and balance of power in the world.

The economic recession existing in almost all Western industrialized nations and the achievements reached by petroleum-exporting countries are events that reflect a radical change in the economic and even political situation and tendencies that had prevailed in the world up to a few years ago. These lead to further reflection on two aspects, to which I attribute great importance:

First, underdeveloped countries should not allow themselves to be led or influenced by intellectuals belonging to the small group of industrial nations. Many may be scientists, great masters of technology, and men and women with a high degree of scholastic achievement, but their theories, reasonings, postulations, and predictions—although often formulated in the best of good faith—will frequently be focused from one perspective only and will undoubtedly be deeply influenced by the fundamental interests of their own nations. Since these nations are a privileged minority, it is most difficult for these thinkers to identify themselves with the genuine interests of the entire international community and even less with the specific interests of Third World countries, the counterpart of the developed world's interests.

An example of opinions circulating in developed nations that, as such, lack any rational or scientific basis is the assertion that the high petroleum prices are the cause of the great economic ailments affecting the world. The high petroleum prices are negatively reflected in the balance of payments and costs at a world level, but the galloping inflation occurring in these nations was developing long before 1973, when the great rise in petroleum prices started.

As well expressed by Dr. Juan Pablo Pérez Alfonzo, founder of OPEC, whom we certainly can include in the small number of thinkers belonging to Third World countries who use their own criteria without being influenced by the intense ideological penetration of industrialized nations, "Those who blame OPEC for the present world crisis are deceiving themselves." He compared the industrial nations' attitude to that of Don Quixote when he assailed the windmills mistaking them for giants.[12]

The sharp increase in petroleum prices has therefore coincided with a serious structural problem that already existed in the great nations of the Western world. Since they developed their economies and consumer societies on the basis of importing low-priced raw products principally from Third World countries, they created a structure of dependency having repercussions today through increased petroleum prices and the embargo measures temporarily adopted by Arab countries on their exports. This type of development, based on ever-increasing massive consumption as well as other factors, has created serious inflationary pressures that have alarmingly affected industrial economies.

It is therefore erroneous to point to petroleum as the cause of the recession in consumer nations. Nor does the phrase "petroleum crisis," so frequently used as a slogan in the daily terminology employed for analyzing world economic problems, apply to reality. If there did exist a "petroleum crisis," it was toward the end of 1973 and beginning 1974, when the Arabs embargoed their oil exports to certain consumer nations, which saw their economies quiver upon lacking petroleum supplies. However, after supplies became normalized, the use of this term seems incorrect. To be consistent industrial nations should already have used the expression "industrial crisis" for many years to describe the inexorable and continued increase in prices of finished products which they themselves export and which contribute to further impoverishing Third World countries.

Naturally, the coining of such phrases as "petroleum crisis," which become unconsciously repeated daily in underdeveloped countries, has a clearly defined aim that unfortunately is not understood by the great majorities, i.e., to make the major petroleum-exporting countries appear to be responsible for all the economic ills burdening the world; to arouse public opinion against them and thereby force a substantial drop in petroleum prices; and eventually, to justify any belligerent action against them.

The foregoing does not, however, disregard the fact that the increase in petroleum prices (on the order of 400 percent during 1973–1974) has contributed to accentuating the inflation prevailing in the world and to aggravating the economic situation of Western developed nations. Experts estimate that the increase in petroleum prices is contributing by

[12]*El Nacional* newspaper, Caracas, Jan. 8, 1975.

less than 3 percent to the inflation existing in those nations. It is also true that the rise in petroleum prices is negatively affecting balance of payments. At 1974 price levels, industrial nations made payments of $22 to $45 billion to OPEC countries. As a consequence of several factors, among them oil, the actual rate of economic growth in Western industrial nations decreased from 6.7 percent recorded in 1973 to only a little over 1 percent as an average in 1974.

On the other hand, we must bear in mind that the economic recession in developed nations directly and negatively affects the economies of nonpetroleum Third World countries. In fact, exports by the latter to developed nations represent approximately 75 percent of their total, and hence a decrease in the growth rate of industrial nations becomes rapidly translated into a decrease in the demand for products exported by emerging countries, in turn causing a decline in the latter's import capacities and thus in their rates of growth.

This situation should be thoroughly considered by petroleum-exporting countries when they formulate their policies and strategies. Such strategies, although primarily oriented toward consolidating their achievements in relation to their income as petroleum exporters, should also be oriented toward inproving the economic and social conditions of Third World countries, especially the poorer, and toward helping them overcome the ties of dependency to which they have been subjected, without unleashing a world economic crisis that could have consequences of a magnitude at present difficult to measure or predict.

FUNDAMENTAL GOALS

Overcoming Dependency

I have analyzed economic dependency in detail because I feel it is essential for all petroleum-exporting and Third World countries to have a precise concept of the situation in order to wisely and soundly orient their line of action to free themselves from these bonds. Venezuela especially, as one of the principal petroleum countries, should formulate a clearly defined strategy.

Achieving this goal should be the fundamental objective of Venezuela and all OPEC and Third World countries; however, a country can become independent only when its people (not merely insincere and demagogic small groups) are united by philosophy and ideals capable of serving as an incentive for such a great effort.

Third World countries should be fully conscious of their failures, weaknesses, and errors; however, they must also be aware of the unfair situation of dependency to which they are subjected. Both their leaders and those of industrial nations must realize that if measures leading to international justice are not taken soon, a world confrontation might occur, not between the great rival blocs represented by conflicting political systems, but between the great number of poor countries and the small group of rich ones.

All industrialized nations, whatever their ideology, must realize the urgent need to contribute to establishing a more equitable status for Third World countries, if they genuinely wish to avoid dangerous, difficult, and even violent situations. Otherwise it may be precisely with this large group of poor countries that a serious confrontation occurs, if over the short term remedial measures are not factually and immediately taken, devoid of fallacies or dogmas.

In any case, it is the dependent countries themselves that must take the initial step toward their own independency, creating an understandably realistic and objective philosophy. Only a thorough awareness of their vital interests, and courageously firm efforts for emerging from underdevelopment, will prevent a continuation of their dependency and close the gap between rich and poor that could otherwise lead to such a dangerous confrontation.

Skeptics and "practical men," concerned only with their personal welfare, will probably fatalistically continue accepting as fact the injustice prevailing in this relationship of dependency. They will allege that the world has always been plagued by inequities and that it is useless and utopian to openly confront the big powers—lacking the necessary weapons—and from whom, furthermore, they receive cooperation and assistance. Many of these practical men were also very skeptical over OPEC's possibilities for success when it was founded. At that time some even derided this new organization, which embraced countries having such varied cultures and social, political, and economic conditions, classifying its objectives as being utopian. However, we have seen the fruits that, in practice, the petroleum-exporting countries are reaping thanks to OPEC.

This skeptical attitude, appearing to be practical and realistic, is basically an attitude of submission and passivity, and those adopting it are accepting, without benefit of analysis or taking stock, certain apparently logical concepts which in truth reflect the interests of industrial nations and which, as we have seen, are fundamentally opposite to those of underdeveloped countries.

In Chapter 5 I referred to posted prices for oil which, outside of the United States, were applied to petroleum-exporting countries and which, up to the end of the 1950s, followed the pattern of high United States prices. This phenomenon has possibly not been analyzed by many leaders of underdeveloped countries. They perhaps have not realized the fact that the United States, when it was a net petroleum exporter, was interested in making sure that prices of other producers—then its competitors—were equal to those of the United States.

However, from 1948, when the United States became a net petroleum importer and was obliged to buy increasing quantities of petroleum every year from abroad, it did not find it convenient to pay high prices for such imported crude and by-products. Since then, posted prices became systematically shifted from the United States to production centers, and posted and closing prices for hydrocarbons proceeding from underdeveloped exporting countries started gradually to decrease, thus departing from the pattern of using United States oil prices as a quotation basis.

In turn, European countries and Japan were able to develop their postwar economy by buying petroleum from the Middle East, Africa, and Venezuela at low prices, while prices for their finished products increased each year. As asserted by Iran's Minister of Finance, Jahangir Amuzegar, in an interesting article entitled "The Oil Story: Facts, Fiction and Fair Play":

> Thanks partly to the uninterrrupted supply of cheap petroleum (as low as $1.25 per barrel toward the end of 1970) Japan and European Economic Community countries raised their industrial production, increased exports, improved their balance of payments, saved their own sources of solid fuels and were able to create a great reserve of funds. However, Western countries are not only unashamed of that enrichment at the expense of their helpless suppliers, but their politicians, economists and petroleum experts have even become indignant over having to pay a fair price for imported petroleum, due to its shortage, through collective negotiations with OPEC countries.[13]

Because of the preceding facts, and because figures clearly show that some developed nations (the United States, for example) are exhausting their own reserves of raw materials needed for their industrial machinery, and because others completely lack such reserves and consequently depend, and will increasingly depend, on those owned by Third World countries, we may conclude that the latter are today in a position to develop clearly defined goals and adequate strategies of truly nationalist content, without regrets or self-recrimination, without verbal aggressions, but strictly oriented to emerging from their status of dependency and securing the prosperity of their economies.

This is the authentic nationalism that should characterize future action

[13]*Foreign Affairs*, July 1973, p. 682.

by underdeveloped countries. This is the true nationalism, as yet not formulated, which should be developed, particularly in Venezuela. This should be the attitude of a country being projected toward a clear process of strong assertion, a country which, supported by ethics capable of creating its own values, is initiating a program of development. Such an attitude and philosophy, together with its goals and strategy, are the only elements that can mobilize the energy of a nation seeking transformation and dynamic and effective development.

I wish to clearly state that the prevailing international social injustice that I have attempted to objectively describe must not exalt emotions in underdeveloped countries to a point of leading them to laments, complaints, or hostile attitudes. They must avoid sinking into a rhetoric of concepts charged with emotion, such as "exploitation," "imperialism," "low income per capita," "ignorance," "illness," "racial discrimination," "lack of assistance and insufficient comprehension by developed nations," "technology gap," and so on. This type of rhetoric might cause deviation from the straightforward and solid path leading to overcoming dependency.

In forums at both national and international levels, representatives of Third World countries usually unburden their logical frustrations. This is very human, but not at all practical. History has demonstrated that it is not possible to overcome injustice merely by accusations or postures favoring morality, equilibrium, and justice.

Even in the cases of injustice committed in international spheres against underdeveloped countries, there are reasons, or rather rationalizations, on behalf of the small group of wealthier nations, aimed at mitigating their sense of blame to demonstrate that the privileges they enjoy are a very fair prize earned through their work capacity, willpower, and tenacity; through their accurate planning, creative skills, discipline, and ability to overcome their own national problems; and even through their armed interventions in defense of freedom and "to preserve the present world order."

The philosophical, ethical, moral, and legal arguments supporting the cause of Third World countries should be clearly, systematically, and objectively analyzed and communicated to their people as much as possible, so that they are fully conscious of the attending reasoning. This knowledge and consciousness will not alone suffice for solving problems, but it should, nevertheless, be the basis for vigorous, pragmatic, and programmatic action. In this way underdeveloped countries—supported by the public, aided by a collective effort and combative attitude, and with superiority in fundamental raw materials—will be able to stand up to dominant nations and achieve world equilibrium in many aspects, especially insofar as international trade relations are concerned.

We have seen from this book how this action has already commenced in petroleum countries, through OPEC. But there is still a long way to go, and the difficulties that must be overcome in this vital industry are great. Undoubtedly, great consumer nations will resort to all their power, not only in their attempt to curb as far as possible the continued increase in petroleum prices, but also in seeking to force a decrease in the prices. Obstacles might possibly also arise for achieving maximum permanent effectiveness from OPEC, due, among other reasons, to differences in certain individual conditions of its members, in their respective petroleum industries and their social and political idiosyncracies.

From the foregoing we can see that the fundamental goal to be pursued by Venezuela, as well as other petroleum-exporting countries and Third World countries in general, is to wholly overcome ties of dependency to the small group of dominant industrial nations. Naturally, this is a very ambitious goal having diverse applications, within both national and international spheres.

From the spiritual viewpoint it requires, as we have seen, overcoming intellectual domination. This is a slow process involving a multiplicity of aspects that cannot be included in this book due to space limitations. From the economic viewpoint, there is one outstanding aspect that I have referred to frequently: the attainment of a greater international trade equilibrium with industrial nations, which requires, among other things, the consolidation of petroleum prices; the strengthening of prices for basic raw materials produced by the other Third World countries, to which effect petroleum-exporting countries must provide their maximum possible collaboration; and the transfer, in adequate terms, of technology from the more advanced countries. To these and other goals I shall refer in the follwing pages.

Consolidating Petroleum Prices

Obviously Venezuela and all petroleum-exporting countries are profoundly interested in consolidating their extraordinary achievement of higher petroleum prices. The remarkable increase experienced in these prices between 1970 and 1974 (especially 1974) is a fair compensation for the depressed prices at which this nonrenewable resource of energy was sold in the past. It is therefore erroneous to limit an analysis of the evolution of petroleum prices to the short period comprised between 1970 and 1974. If an analysis is thus limited, we must logically conclude that these increases not only surpassed the most optimistic predictions, but also were perceptibly higher over that short four-year period than the increase in prices for the developed nations' finished products, which followed only a moderate tendency compared to petroleum prices.

However, this comparison changes radically if we take into account, as we should, longer periods of time. Upon studying the chronological series applicable to Venezuela, and using the twenty-year period between 1950 and 1970, we can see that the value of Venezuela's imports (proceeding mainly from the United States, and in lesser proportion from Western Europe) exceeded approximately $533 million in 1950, and were close to $17 billion in 1970. Taking these figures in terms of units, the price in the two years at each extreme of this series increased from $310 to $421 per imported metric ton.

On the other hand, Venezuelan exports of petroleum and by-products (also destined mainly to the United States and secondly to Western Europe) had a unit value of $15 per ton in 1950, and decreased to $13 per ton in 1970. In other words, in that same twenty-year period, the unit value of petroleum and derivative exports decreased by $2 per ton, whereas imports of finished products increased by $111 per ton. Furthermore, inflation is so high in the world and in great consumer nations that the nominal increase in petroleum prices recently experienced is in actual terms much less.

Likewise, it is well known that the international petroleum companies, which up to the present have been extracting petroleum in underdeveloped countries and which still control international trade, took advantage of this situation by disproportionately increasing prices in detriment to consumers. Therefore, although fiscal income in Venezuela and in other export countries increased as a result of the new petroleum reference prices, the profits of most of the international petroleum companies also increased. In the first nine months of 1974, the combined profits received by Exxon, Texaco, Mobil, Standard Oil of California, and Gulf, for example, increased by 50 percent, compared to the same period in 1973.

The reaction of governments of import countries, by impeding the companies from further transferring to consumers the taxes charged by OPEC, has managed to slow down the growth of company profits. In fact, profits obtained by a group of United States companies, including the five mentioned previously, which had registered an increase of 103 percent in the first half of 1974 compared to the like period in 1973, decreased by nearly half (50 percent) when the results of the third quarter of 1974 were added.

These reasons justify what Venezuela and other OPEC countries must pursue as a fundamental goal: maintaining the 1974 level of income per barrel. This means that reference prices should be periodically adjusted to compensate for loss in currency value of income received, owing to inflation.

Nevertheless, OPEC countries must be alert to the evolution of economic conditions in great consumer nations as well as underdeveloped countries, to provide the latter with all possible cooperation and adopt a flexible position in accordance with the future evolution of events, so as to avoid grave economic recessions while a more balanced world economic order becomes established.

To achieve this fundamental goal, leaders of petroleum-exporting countries, through OPEC and individually, must employ action fitting of great statesmen. They have no other alternative!

Appropriately Investing Petroleum Resources

The increased revenue being received by OPEC countries through higher petroleum prices is generating excess funds, which according to World Bank estimates reached approximately $80 billion in 1974 and is estimated to reach $100 billion in 1975. The same source indicates that these excess funds might reach approximately $173 billion and $256 billion by 1980 and 1985, respectively. In these forecasts, the Bank has not taken into consideration any substantial changes in price levels or in the pattern of demand for petroleum.

The World Bank has also made some studies of the international monetary reserves in possession of OPEC members, as follows: between December 1973 and September 1974, such reserves increased from $14.5 billion to $38.4 billion; in 1980 they could amount to approximately $653 billion, and by 1985 nearly double ($1.2 trillion).

It therefore seems obvious that export countries already have, and will further accumulate in forthcoming years, vast excess funds that cannot be invested locally in high proportion, at least in the first years, because OPEC members—some less than others—lack sufficient cultural, social, and economic development to adequately absorb this oil income, which in less than one decade has increased fourfold.

This book cannot, even where Venezuela is concerned, possibly estimate the amount of oil revenue that will be accumulated by OPEC countries in the next ten years, or its effect on the balance of payments or international reserves, or the proportion of such income that could be invested locally or abroad. These estimates, logically, must be made by each petroleum-exporting country and checked among them for the purpose of ascertaining the accumulated combined figure and the world impact it could have.

I might mention, however, that any estimate of these funds must take into account the following basic aspects: the possible evolution of the market of each country in the period under consideration; respective productive capacities, which in turn are the result of current reserves and of those that can be developed in the future; the policy that, it is hoped, will be followed by the countries individually—and also jointly—in relation to production and exportation; the estimated development that might be achieved by great consumer nations for alternative sources of energy; the perspectives of world petroleum demand; and finally, the price evolution that might be expected, which in turn is subject to the preceding factors.

Future price levels will also logically be a consequence of another series of factors, such as the functioning of OPEC in future years, which in turn will depend on the support that its members give its activities. Thoroughly detailed projections based on expected future petroleum income should be carried out by Venezuela and each OPEC member country, and revised regularly. On such bases, Venezuela and the other countries must execute their national plans, including investment budgets, both domestic and foreign.

Local Investments

It seems obvious that another fundamental goal of Venezuela and other OPEC countries should be to invest with maximum possible efficiency the major portion of these vast funds in local projects and services of an economic and social character that will enable them to achieve a high degree of development in the shortest possible time.

In order to achieve this basic goal, petroleum countries, taking into account their human and technical resources, should intensify the preparation and execution of economic projects in industry, agriculture, livestock, and so on, based on carefully studied medium- and long-term plans and programs. The plans should be ambitious but very realistic, to avoid the risk of squandering great sums of money in projects that may prove to be failures. The programs must also be subject to strict supervision, evaluation, and periodic adjustments, concordant with practical fulfillment and changing circumstances.

Local investments should, of course, be oriented to accelerating the development and expansion of export countries. These countries, like Venezuela, should not waste this unique opportunity to achieve, over a short time, a high level of development in all fields. For this reason the elaboration of these medium- and long-term development programs is now especially important, to clearly establish the desired national goals.

Not only is it necessary for these goals of national development to be adequately formulated and set in motion, but it is essential that they be efficiently carried out so that the entire population—not only its leaders— may feel united in the ideal of fulfillment. This should be the task of governors who, through channels of communication and every means at their disposal, should contribute to creating a national moral wish for improvement and development, awakening the public to the fact that these extraordinary revenues are temporary and should be taken advantage of with utmost urgency, and as efficiently as possible, for the benefit of present generations and especially future generations, since the latter may not have such income proceeding directly from petroleum.

This fundamental goal cannot be achieved by all OPEC countries under similar circumstances of time and execution, because apart from the fact that the amount of income received varies according to production, different conditions also prevail in their workforce levels of education, present economic development, social aspects, managerial and technical capacity, etc. I shall not analyze these differences, but as an illustration, and not fully identifying myself with these concepts, I shall briefly refer to the classification of OPEC countries made by Galeazzo Santius, the renowned Italian expert:

1. Countries having scarce population: Kuwait, the Persian Gulf emirates, and Libya. A parallel increase in imports cannot correspond to the tremendous petroleum income. Consequently, a great volume of excess funds will be available. Between 1973 and 1978 a 15.4 percent and 20 percent increase in imports is foreseen in Kuwait and Libya, respectively.

2. Saudi Arabia, with 8 million inhabitants, has vast reserves of copper, iron, and silver, besides petroleum; but it has insufficient human and technical resources to develop them. Nevertheless, in the period 1973–1978, its imports will increase by 47 percent, and a great excess of capital will also be available for capital movement.

3. Countries which have development plans for rapid execution and which are equipped with the basic human and technical infrastructures: Iran, Venezuela, Algeria, and Iraq. These countries will be in a position to absorb most of their petroleum income. For the period 1973–1978 an annual increase in imports is foreseen, amounting to 40.5 percent in Iran; 30.5 percent in Venezuela; 15.7 percent in Algeria; and 42 percent in Iraq.

4. Densely populated countries: Indonesia and Nigeria. Having a low level of development, their capacity for using the oil money will be restricted by their rudimentary economic structures. An annual increase of 24 percent in imports is foreseen in Indonesia and 20 percent in Nigeria.

In the specific case of Venezuela, it is important to point out that numerous investment projects exist and are in the development stage, especially of an industrial and infrastructure nature, which will absorb a considerably high proportion of its new petroleum funds. Consequently, the government is making an inventory of all such investments and of their capital requirements, so as to formulate a clearly defined foreign investment policy based, among other factors, on Venezuela's needs for liquid resources to finance its domestic investments.

As an example of such investments, I might mention the Corporación Venezolana de Guayana (CVG), a state company having as an objective the complete development of the Guayana region—which is the vast territory situated in the southeastern area of Venezuela, possessing voluminous mineral and hydraulic resources and able to house giant industrial complexes. CVG operates as a holding corporation, with several wholly-owned subsidiaries, such as Siderúrgica del Orinoco (SIDOR) and the Electrificación del Caroní (EDELCA).

CVG has projects ready for execution between 1975 and 1979, among which are the expansion of SIDOR and of the hydroelectric complex and the manufacture of motors, tractors, paper, pulp, chemical products, and industrial gas. These plans will require investments of approximately $5.3 billion and will include the expansion of existing companies or the formation of new ones, in which CVG will contribute majority capital or any other proportional amount.

Besides the capital contributions by CVG in each of the companies within this vast program, the execution of the projects will require additional funds, which could be financed with petrodollars through the Venezuelan Investment Fund (to which I shall refer later) according to the following annual estimates:

Year	$ (Millions)
1975	3,839
1976	877
1977	1,017
1978	783
1979	633
Total:	$7,149 or $7.1 billion

These figures correspond to direct investments in industrial projects and therefore do not include—except for the hydroelectric development—the necessary infrastructure development for covering the needs of the vast Guayana region's accelerated growth.

The development of this infrastructure will logically also absorb important quantities of the Venezuelan Investment Fund's petrodollars.

To illustrate the magnitude of the hydroelectric project, the great Guri Dam (River Caroní, Bolívar state) generates 985,000 kilowatts per hour with five units already installed and in operation, and by mid-1975 a sixth unit will be operating, adding 240,000 kilowatts per hour. The expansion projects include the starting up by 1979 of four new units, having a generating capacity of 1,600,000 kilowatts per hour; later, and to complete the thirteen-year program up to 1987, another ten units of 600,000 kilowatts per hour each will be incorporated (the greatest at present manufactured), for an additional 6,000,000 kilowatts per hour. In this manner total installed capacity will be approximately 9,000,000 kilowatts per hour by 1987, with the Guri Dam thereby possibly becoming the largest in the world.

Energy generated will be used throughout all Venezuela by means of existing interconnecting networks and others to be built in the future.

Investments Abroad

The amount by which Venezuela's total state income exceeds local investment and expenditure foreseen for each year should be allocated abroad. This is such an important, complex, and intricate task that the government should establish clearly defined plans concordant with available funds, and create effective tools for carrying out such investments. For that purpose, a study of these excess funds is presently being made, and a special organization has been created, the Venezuelan Investment Fund.

The Venezuelan Investment Fund.   The creation of the Venezuelan Investment Fund, under decree dated June 11, 1974, was an excellent measure adopted by Venezuela. This organization, having a legal status and autonomous patrimony independent from the National Treasury, is a dependency of the Presidency of the Republic. One of its priorities is to prevent the totality of petroleum income from entering into circulation and thereby causing disadjustments in the country's national economy through its mere volume.

To that effect, the Venezuelan government has established that 50 percent of income proceeding from petroleum and gas exploitation taxes, as well as the income tax from these activities, shall become part of the patrimony of the Venezuelan Investment Fund, as soon as collected. The fund thus created is, as mentioned, independent from resources allocated the nation's general budget of income and public expenditures.

In 1974, the funds allotted amounted to $3.09 billion, and $3.15 billion have been budgeted for fiscal 1975; by December 1975 its capital should amount to $6.2 billion.

The principal aims of the Venezuelan Investment Fund are to complement the financing of the expansion and diversification of the nation's economic structure; to make income-producing investments and placements abroad; and to develop programs of international financial cooperation.

To attain these objectives, the Investment Fund is empowered to finance the foreign components of local industrial projects and projects abroad, with participation of Venezuelan capital; to complement the financing of major investment projects covering agriculture, manufacture, and exportation; to acquire shares and participation in companies of foreign or mixed capital established within the country; to participate in companies and financings abroad, and so on. In general, it might finance or invest, directly or indirectly, through financial or associated institutions, through public or private international correspondents, or through international public institutions, real estate and securities of ready marketability in freely exchanged currency.

To efficiently develop its operations, the Venezuelan Investment Fund is empowered to enter into contracts for banking correspondents, consultants, trust management, and others of a similar nature.

Liquid assets not invested by the Investment Fund shall be maintained in deposits in first-class banks abroad, and in foreign public securities having stable markets permitting immediate sale in freely exchanged currencies. On the other hand, to avoid hasty liquidation of a portfolio under unfavorable circumstances, or to advance operations anticipated in its investment programs to take advantage of market conditions, the Fund is entitled to contract foreign or domestic credits, or issue promissory notes for terms not exceeding one year and in amounts not higher than the foreseeable fiscal contributions for the following fiscal year.

The Investment Fund is likewise entitled to buy from, and sell to, the Central Bank of Venezuela all type of documents and securities in foreign currency that are eligible for forming a part of the latter's portfolio in accordance with legislation in force.

The Fund's main body of operation is the General Assembly, composed of fifteen members, i.e., eight ministers representing the executive branch; two representatives from Congress; the president of the executive board; the president of the Central Bank; and one representative each from the private banking sector, the labor sector, and the private trade and production sector. The Investment Fund is administered by an executive board composed of five members, named by the President of the Republic, with one member designated as president of the board, who is in direct charge of the administration of the Fund's operations.

Achieving More Equitable International Trade Terms

From the preceding considerations and my comments relating to the need to overcome the ties of dependency, it can be seen that another fundamental goal lies in the establishment of more equitable international trade terms.

Petroleum-exporting countries have now achieved higher prices, which are beginning to compensate for the subsidy that we might say they have been granting—up to a short time ago—to consumer nations by means of the depressed prices they formerly received for their petroleum. This achievement alone is already an important step forward. However, it is also necessary for these countries to obtain favorable treatment in the future in prices of finished products exported by the great consumer nations.

In other words, OPEC countries, to be able to attain one of their greatest economic goals—large-scale industrialization—must secure free access to the markets of the great industrial nations. They must request those nations to eliminate artificial import barriers against finished products manufactured in OPEC countries, which should be subject only to the limitations proper to the economic laws of supply and demand.

This objective is now acquiring greater standing in view of the United States Trade Act of 1974, which follows the traditional line of placing customs barriers on products, raw or manufactured, proceeding from other countries. This Trade Act, passed by the U.S. Congress on December 20, 1974, and signed into law on January 3, 1975, has a marked retaliatory nature against petroleum-exporting countries and against any other policy considered by the United States Executive to be detrimental to United States interests. In effect, it empowers the President to judge and qualify the trading and economic policy of every country, or group of countries, exporting to the United States, in order to decide whether to grant or deny the preferences contemplated in the law.

Countries denying the United States fair and reasonable access to their markets and to their sources of raw materials will not benefit from the advantages granted by the law. The decision concerning the eligibility of developing countries for benefiting from these advantages will be made by the United States government, which will judge whether access allowed the United States to the sources of raw materials is "fair and resonable." It is also indicated that no benefits will be granted to any emerging country should the country concerned be a member of OPEC.

In addition to limitations toward countries that might be favored by the law, an attempt is made to cause difficulties for associations that underdeveloped countries might form in the future. In effect, the law denies preferences to those exporter countries that provoke market imbalances through production restrictions or through price increases. Thus, a further attempt is being made to restrict the freedom of producers and exporters of raw materials to become united and form a bloc for protecting their own economic interests, such as OPEC, which has been so highly successful.

This United States legislative measure, permitting its government to suppress certain trade advantages, may cause a division among Third World countries, between those accepting United States "benevolence," submitting to its rules of the game, and those raw-material-producing countries that are willing to vigorously defend their true interests as exporters to achieve a more equitable distribution of world wealth.

Because OPEC countries must strive for greater international economic justice, they must use the power of negotiation and persuasion now on their side, since they possess more than two-thirds of the world petroleum resources, for achieving these objectives both for their own benefit and for the benefit of Third World countries.

Acquiring Adequate Transfer of Technology

Industrial nations—through the effort and imagination of their people, specializing in industry, agriculture, and varied types of services, with cooperation from centers of higher learning and government sectors— have created a wide range of technology. This is obviously the case in the United States; abundant literature has been written concerning the technological development achieved in the last thirty years, allowing it to attain a degree of advancement unknown before in the history of mankind, infinitely superior to that achieved throughout all the preceding years of civilized life. This undeniably immense technological treasure accumulated in developed nations, covering the most diverse facets of modern life, is invaluable, even in monetary terms.

On the other hand, petroleum-exporting countries today have the financial resources necessary for acquiring the technology they need, and also the negotiating power for attaining terms of technological transfer under more equal conditions than those previously prevailing, which have been one of the most important factors in the drainage of the Third World countries' balance of payments.

To obtain ample technological transfer under fair terms, contributing to rapidly developing their economies, and in general, to achieving a high degree of development at all levels, is therefore one of the most important goals that should, and can, be reached by petroleum-exporting countries.

THE BASIC STRATEGY

It is logical that well-conceived plans alone are not sufficient; it is essential for them to be wisely and fully executed. To that effect, a fundamental requisite is the employment of a series of strategies and tactics created with maximum refinement and regularly reviewed, so as to be able to adapt them to the dynamics of events which, especially in the petroleum world, are so changeable. If goals are subject to change in accordance with the evolution of events, this principle has even greater application to strategy and tactics, which are the instruments for achieving them.

It is thus that changes in strategy and tactics must be made with the highest degree of rapidity, in order to swiftly adapt to new circumstances. This occurs, for example, when a strategy is outlined for a given situation and then the opposite side changes its position in facing the same situation. Logically, a revision and reformulation of the initial strategy immediately becomes essential. It is therefore convenient to have constant knowledge of the goals, objectives, strategies, and tactics of the individuals, entities, or countries with which there exists a conflict of interests.

Although it is beyond the scope of this book to analyze in detail the fascinating subject concerning the strategies that Venezuela, jointly with OPEC countries, could develop and pursue, I cannot resist briefly referring to some that I believe are extremely important at present for achieving the objectives outlined.

The Strategy related to Petroleum Prices

In order to maintain and to consolidate petroleum prices, an essential condition is the increasingly closer relationship among OPEC countries and among these and all other underdeveloped countries. I have frequently referred to OPEC, which was in great measure created through Venezuela's initiative and which—contrary to the opinion of a great number of its detractors—has been acquiring strength and cohesion and achieved, after ten years, what possibly might be classified as the most outstanding conquest of freedom in the economic field ever reached by emerging countries.

Essentially, this achievement consists in having received fair petroleum prices; in having broken the ties represented by the control over oil resources exercised by the great international companies and by the great powers; and even, in some cases, in having pressured these powers in order to achieve certain political ends. These achievements by OPEC countries are also contributing to creating a new world economic order, and thereby represent a wide-ranging peaceful revolution, headed by members of that organization.

For practical reasons, I feel it advisable—in order to formulate suggestions related to the strategy that should be adopted by OPEC members for consolidating and advancing their achievements, especially in relation to prices—to first review the strategy and principal tactics being followed by the United States and other industrial Western nations to counteract or neutralize these attainments as far as possible.

The Strategy of Petroleum-Consumer Countries

The Western nations have designed and put into practice a strategy that, in its fundamental aspects, consists in weakening OPEC's front through various tactics devised to decrease oil prices. The principal ones are:

1. To reduce petroleum consumption. This tactic (apart from praiseworthy aims, such as avoiding past excessive consumption and waste)

would seek to create a production surplus in export countries in an attempt to force them to reduce oil prices by trying to maintain a high rate of production. Initially, Henry Kissinger had proposed that the great industrial nations decrease consumption by 3 million barrels per day by the end of 1975. However, other proposals have since emerged, one of the most outstanding appearing in the report of January 11, 1975, by the U.S. Senate's Subcommittee on Foreign Relations, concerning multinational corporations.

Democratic Senator Frank Church, who presides over the subcommittee, told the press at that time that a decrease by 3 million barrels per day on behalf of the principal industrial nations "is insufficient" to force OPEC to significantly reduce prices, and that the United States should "attempt (an action) to dissolve the cartel" by creating a world surplus of crude oil; consequently, a decrease in consumption on the order of 6 million barrels per day was proposed. Senator Church asserted that in such a case, excess production would be sufficient to avoid OPEC's being able to prorate production decreases and income among its members. He furthermore added that OPEC members had never been successful in their prorating endeavors.

Evidently, the tactic directed toward decreasing consumption to provoke excess production has been proposed not only by the United States executive branch, but also by its legislative branch.

2. To create a bloc of industrial nations. I have already referred to this tactic, propitiated principally by the United States, to strengthen negotiating power in confronting OPEC. We have also seen that so far attempts to definitely form this bloc have not materialized. Besides achieving other objectives—such as creating a petroleum reserve for emergencies—this bloc of great consumer nations would also have the main aim of forcing a reduction in petroleum prices.

3. To divide Third World countries into petroleum producers and nonproducers, based on the greater burden weighing on the latter due to the substantial increase in petroleum prices. This is a tactic that leaders of great consumer nations have openly supported.

4. To threaten military intervention. As defined on one occasion by the Russian Tass Agency, this tactic constitutes a cold war expedient. It is reminiscent of that undertaken by the great powers in the past. OPEC countries have little experience in this type of warfare and must rapidly acquire that experience in order to face it effectively, carefully elaborating whatever strategies are applicable.

On the other hand, there is a risk that threats by some of the great powers might materialize, becoming a belligerent act in fact and deed. OPEC members should be fully prepared for such an extreme situation, and I shall later refer to some fundamental aspects in this connection.

5. To exercise other pressures, such as that represented by the new United States Trade Law.

Another not very apparent objective of the strategy being employed by Western industrial nations, through the aforementioned tactics, consists in preventing a future rise in oil prices—in other words, keeping OPEC members from making any further price increases. I might say that this last objective—at least for the time being—appears to have been achieved by the great petroleum consumers. In effect, OPEC countries—apart from some tax increases assessed on the oil companies—do not appear inclined to make any further increases in oil prices, as in the recent past.

In this respect, possibly in the near future, a process of "re-flation" might occur, to a greater or lesser degree, in each industrial nation. In other words, the severe measures being adopted to control, curb, and decrease inflation would be lifted and a new inflationary resurgence would occur. Some symptoms of this are already in evidence in some European countries and the United States (for example, interest rates have begun decreasing). Should this phenomenon of "re-flation" occur, and should the present level of petroleum prices be maintained—or should they increase, but not in proportion to the indexes of inflation in industrial nations—then they would, in relative terms, be decreasing. This certainly is an aspect that should be carefully studied by petroleum-exporting countries.

In any event, OPEC countries should closely follow the attitude, strategy, and tactics adopted in the future by certain industrial nations (such as Japan and the European Common Market countries) that are seriously affected by the increase in petroleum prices. For example, they should be alert to the possibility—in the event of a new armed conflict between Arab states and Israel (which would be fighting for its survival)—of industrial nations' effecting massive supplies of arms to Israel.

The Counterstrategy

Having outlined the basic aspects of the strategy being used by industrial nations toward OPEC, I shall now mention five points that I believe should be included in the petroleum countries' counterstrategy.

In the first place, OPEC countries should remain adamant in facing the series of pressures being exercised against them by the industrial nations. In counteracting adverse propaganda in its varied forms, OPEC members should make an extra effort to assure that the public in both Third World countries and industrial nations fully comprehends the true situation and realizes that petroleum-exporting countries are not, by having increased oil prices, the principal parties responsible for the inflation and problems affecting most of the world today.[14]

They should endeavor to make the public understand the injustice that has predominated in world trade relations throughout centuries, and especially in this one: the fact that the substantial transfer of wealth shifting today from the richer nations on earth to the group of petroleum-exporting countries, and through these possibly to other Third World countries, is a phenomenon favoring humanity by finally leading to a more equitable distribution of world wealth. I believe these chapters contain arguments to illustrate the reasoning that should be fully divulged throughout the world to attain better public comprehension and to neutralize the negative propaganda released by consumer nations.

In the second place, concerning the decrease in oil consumption by industrial nations, I feel such a measure is healthy and convenient, because to date the industrialized world has been overconsuming—and even squandering—this precious nonrenewable energy resource. However, as such a decrease in consumption is also a tactic being used for creating great surplus and lowering oil prices, OPEC countries should make an in-depth study of this subject and be prepared to rapidly and efficiently meet the situation, prorating decreases in production when necessary.

It is true that OPEC so far has not carried out complete programs of this type; however, this does not mean that OPEC—which has always been ready to face a series of problems and adopt appropriate and practical decisions—cannot do so, or does not wish to do so, especially if thereby it is possible to maintain and consolidate the genuine income per barrel already achieved. Nor must we forget that the volume of income being received by OPEC countries is so vast that they may perfectly well be able to reduce, at any given moment, their total general income in order to maintain the real income per barrel, without affecting their national and even international development plans thereby.

In the third place, in reference to the military intervention that United States spokesmen say would be justified in the event of a strangulation of the industrialized world (the time of its occurrence to be decided unilaterally by developed nations), OPEC members should remain on the alert and take strong precautionary measures. Among these measures could be the official announcement that an armed aggression against any OPEC member would be taken as an aggression against all members and that in this event all petroleum supplies to the great Western markets would be completely cut off.

If a solid and firm position in this direction is adopted by each and every

[14]An example of this tactic for counteracting adverse propaganda might be found in the joint declaration issued in January 1975 by the Shah of Iran and President Sadat of Egypt, with reasoning similar that to expressed herein.

OPEC country, it seems improbable that any power would be willing to launch such a venture that would imply having to unilaterally control a large number of countries throughout the world. This would be a situation even more unendurable for the invading powers should the invaded countries be—quite possibly—supported by the great majority of their Third World colleagues and also by the Soviet Union.

In the fourth place, we must not forget that although it might be asserted that the cold war between the United States and the Soviet Union and satellite countries has in great measure ceased, as a consequence of the political détente initiated during the Nixon administration and the trade interdependency between both blocs, in truth that détente and that interdependency are at present embryonic, fragile, and open to serious risks in this political game of world chess.

Generally, the Soviet Union has until now supported the Arab countries, whereas the United States has supported Israel; it is therefore doubtful that the Soviet Union will refrain from intervening if an armed intervention by the United States and/or its Western European allies takes place in the Middle East.

In this sway of international forces, special thought must be given to the role that might, at any given moment, be played by the People's Republic of China. In many political and economic circles, especially in Western nations, it is the general belief that the true conflict is not really between the United States and Russia, but rather between Russia and Communist China. Based on this argument, there are those who maintain that should any belligerent aggression on Arab petroleum-producing countries occur, Russia would not interfere. It has even been affirmed that secret pacts exist between the world's two superpowers, which would lead to their mutual support in the event of an armed intervention in the Middle East and even to their splitting the dominion over the most important petroleum-producing countries between them.

I feel, however, that this opinion reflects the viewpoints of certain sectors in the Western industrial world and that it is most doubtful that the Soviet Union—which in any case does not depend on petroleum supplies from abroad—would decide to radically change its policy of ideological infiltration for achieving support of Third World countries (including petroleum exporters) by becoming allied to the United States or other Western industrial nations in the venture of invading Middle East petroleum countries. In this policy of ideological infiltration—designed to draw underdeveloped countries' support and favor—the Soviet Union has traditionally competed with Communist China, and there are reasons to assume that this competition will continue in the future.

Within this world context, OPEC members should contemplate in their strategy the eventuality that Communist China—a possessor of the atomic bomb—might also support export countries in the extreme event of an armed intervention by the United States or its allies. In this respect, we must not forget the vigorous position taken by China at the United Nations General Assembly in October 1974, when it voiced its full support of OPEC's petroleum policy. This could be but another circumstance that would prevent an alliance between Russia and the West for entering into armed aggression against the Persian Gulf export countries.

Bearing these aspects in mind, OPEC members should develop a careful strategy toward the bloc formed by the Soviet Union and countries under its sphere of influence, and eventually toward Communist China, so as to obtain assurances that they would represent a balancing factor in the eventuality of an armed conflict with the Western powers.

If the United States, on its own initiative, has in many ways sought an approach, both politically and tradewise, with the Soviet Union and Communist China, it seems absurd to think that OPEC members—who are being publicly threatened by certain Western powers—should abstain from seeking closer association with Russia, its satellites, and Communist China, with a view to attaining an equilibrium of international forces and to preventing the eventual materialization of those threats.

Because of these and other reasons, the recent decision by Venezuela's government to resume diplomatic relations with Cuba (which had been interrupted for many years) makes sense, as well as Venezuela's proposal—although not yet successful—to suspend the sanctions imposed by the Organization of American States (OAS) on Cuba since 1964.

Within all these vitally important aspects, which OPEC should study and then use to carefully formulate strategy, special attention should be paid to the possible risk of a new armed conflict between Israel and the Arab states. Such a conflict would not be—as it has not essentially been in the past—a consequence of existing interests concerning petroleum. The problem is highly complex and for obvious reasons cannot be analyzed here. However, I must repeat that such a conflict could be taken advantage of by Western powers for intervening indirectly in the producer countries of the Middle East. Such a war, apparently regional, could be greatly influential to the petroleum world and could eventually be detrimental to the achievements of OPEC, which must be preserved. Thus, independent of the nonpetroleum aspects that are of such great interest to the Arab countries, the problem should be placed in focus—and strategies formulated.

In the fifth place, I shall consider the possibilities for strengthening the military power of OPEC countries.

I lack eloquence to sufficiently emphasize the importance for militarized countries of the world, namely the great powers, to effect an intensified program for reducing war materials, and for underdeveloped countries to curb any tendency toward increasing theirs. The number of problems pending solution in underdeveloped countries is so overwhelming, and their resources are so scarce (with the exception of the petroleum countries), that it would even be inhuman on their part to divert funds for acquiring war materials.

It is regrettable to see how great powers nevertheless continue increasing their arsenals. Regardless of all the vast publicity and propaganda covering summit discussions between the Soviet Union and the United States directed toward curbing and reducing their nuclear arms race, both powers have been increasing their capacity for warfare. Further, great Western nations—as seen in the case of France and England—in order to increase their exports and favor their economies, constantly promote sales of weapons to underdeveloped countries, many of which are immersed in border wars, or being dragged along by the cold war activities of the Eastern and Western rival powers even today.

Since the objectives accomplished by OPEC have a fundamentally important world projection, they must be preserved through all possible means; and since the great consumer nations have mentioned the possibility of armed intervention, the export countries might have no other alternative, despite the aforementioned principles (which must not thereby lose their importance or stop being humanity's fundamental objective), than to study the possibility of creating their own military force.

Everyone realizes that this is an extremely delicate matter which should be carefully studied. It might well be that the effort implied for a military capacity sufficient to prevent military intervention by great powers is beyond the economic, technical, and administrative possibilities of petroleum countries. But as long as these grave risks persist, which could tumble OPEC's achievements and precipitate a world disaster of unpredictable proportions, petroleum-exporting countries have no alternative to making in-depth studies of plans to expand their armaments, including nuclear.

I am venturing this affirmation because I believe the risks to humanity, should the successful efforts made by OPEC become frustrated in its effort to achieve a more balanced world economy and distribution of wealth, are greater than those that might arise from a certain degree of potentially belligerent defense on behalf of OPEC countries.

All these tactics might possibly contribute to exerting pressure on the Western powers to abandon their policy of confrontation. In exchange, OPEC countries must give them guarantees of stable supply, reasonable regulation of prices (for example, agreeing to avoid any abrupt increase that might go above inflationary growth), and cooperation—when necessary—in financing their petroleum imports.

On the other hand, it seems obvious that to carry out the vast development plans that OPEC countries must undertake, substantial imports of equipment, machinery, and services from industrial nations will be essential, as is already taking place with France and Italy. I wish to stress that such programs and agreements may be realized only in an atmosphere of peace and mutual understanding, never in an atmosphere of belligerence and mutual threat.

The General Strategy

Having schematically outlined the strategy and tactics that are being pursued by industrial nations in confronting OPEC, and also OPEC's counterstrategy, I shall now refer to the fundamental aspects of the general strategy which, in my opinion, should be followed by Venezuela, and possibly by its OPEC colleagues, to achieve the basic goals that I have described in this chapter.

In order to confront present and future great pitfalls and risks, and to efficiently perform the important tasks corresponding to each OPEC member country, these countries must further tighten the bonds uniting them. The successes achieved by OPEC should lead to optimism as to what might be achieved in the future. In effect, notwithstanding the important differences of a political, economic, social, and religious nature among OPEC members, and the important differences in their petroleum deposits and production capacities, OPEC members have been able to maintain a united front that has enabled them to employ a uniform policy in the face of such fundamental decisions as that concerning petroleum prices. Thus the pressures exerted by great consumer nations aimed at obtaining a decrease in prices and at creating a split among OPEC members have, up to the present, failed.

Within the strategy to be adopted in future, oriented toward intensifying the unity between OPEC members, I feel that Venezuela and all OPEC members should widen the horizons of the organization in order to achieve joint action. They should enter into joint programs designed to massively recycle petrodollars toward the Third World nonpetroleum countries.

In such recycling, consideration might be given to four types of basic programs: investment in economic projects of common interest (industrial, agricultural, service, etc.); financing petroleum imports; collaboration in programs tending to regulate and reorient international trade of raw materials; and programs of social assistance to those countries.

Joint Investments in Eminently Economic Projects

I am convinced that the projects which could be jointly executed by export countries in other Third World countries, preferably coordinated by OPEC, are numerous. In many cases these projects could have great breadth. Such joint projects could be oriented toward several fields requiring great investments, such as activities related to the petroleum industry proper, i.e., petrochemicals; development of new sources of energy; the liquefied gas industry; siderurgical, chemical, and transportation projects, etc. In this manner, highly important projects at a world level could be established in nonpetroleum underdeveloped countries.

Such projects, if carried out efficiently, in addition to bringing the petroleum investor countries the corresponding capital benefits, would also intensively propel the economic development of the receiver countries and assist them in attaining diplomatic, strategic, and eventually military support in facing the great powers. At the same time, the fundamental and strategic goal of strengthening the bonds of unity among OPEC members would be achieved, thereby further consolidating OPEC.

Many export countries, some individually and others jointly, are already studying and executing programs of this nature. I might cite, for example, the project that Venezuela is studying for establishing a refinery in Costa Rica; to that effect it is seeking participation by Ecuador and by Persian Gulf countries.

I realize that this highly ambitious strategic action is a difficult task. The formulation of the projects, their study, and their final execution are extremely complex because of the vastness involved, the lack of administrative and technical resources, and the multiplicity of parties and interests involved. However, the moment is so opportune, and the advantages are so obvious, that OPEC members should make every possible effort to assure that these projects become multiplied and are carried out satisfactorily.

Venezuela, although having petroleum resources substantially inferior to those of other OPEC members, such as Saudi Arabia and Kuwait, possesses a higher degree of economic, social, and political development that might be taken advantage of to the benefit of all. Therefore, the Venezuelan government should, through its Investment Fund and through other means (including those in political and private sectors), take the initiative to discuss possible ventures with its OPEC partners and formulate concrete recommendations and proposals in this respect. This is one of the reasons for accelerating and expanding the technical and administrative organization of the Venezuelan Investment Fund as soon as possible.

In nonpetroleum emerging countries there exist numerous investment opportunities which, if well formulated and studied, could prove to be most profitable both to the receiver countries and to the investor countries. We must not overlook the fact that, in general, Third World countries are a reservoir of natural resources. The exploitation of these resources could be accelerated and expanded with decided collaboration by petroleum-exporting countries, with beneficial results to humanity.

It is not therefore true, as asserted by spokesmen of Western nations and the world press, that Third World countries are not in a position to absorb an important proportion of petrodollars in interesting projects. Such funds are at present oriented mainly toward institutions in industrial nations. It is true that the development of such projects requires a technical capacity lacking in underdeveloped countries; however, this lack of technical know-how might be overcome by petroleum countries through contracting the necessary experts and technicians from various parts of the world.

As inevitably a certain period of time must elapse for the development of programs such as these in Third World countries, the corresponding funds could continue to be invested temporarily, in high proportion and over the short term, in industrial nations, and above all in other programs such as those that I shall now describe.

Financing Petroleum Imports

Obviously Third World countries, especially those having lesser resources, will need financing to meet their increased petroleum-importing accounts. To that effect, OPEC members should establish normal financing programs.

It is estimated that petroleum-exporting countries had a surplus of $60 billion to $80 billion in their 1974 balance of payments. This surplus, of course, will originate as a counterpart, deficits, in all petroleum-importing countries. In underdeveloped countries, the problem will be more acute, especially taking into account the substantial increase in prices of other essential products, such as grains and fertilizers. In great industrial consumer nations, the problem will be of a much higher degree, although less acute because of the multiplicity of their resources. It is estimated that the latter's deficit will have oscillated between $50 billion and $65 billion in 1974.

The situation in 1975, and that of future years, might seriously affect the stability of the international economy, for which reason petroleum-exporting countries must contribute to financing the petroleum deficits. Studies should be regularly and thoroughly made by OPEC to coordinate measures that its members might adopt in this respect.

In principle, it has been considered that a traditional body, such as the International Monetary Fund, could at least partly handle these functions, and IMF has already received petrodollars for that purpose. Although in the general program preference should be given to Third World countries, and those having the least economic possibilities, consideration might also be given to financing imports of the great consumer nations as it gradually becomes perfected.

On January 11, 1975, representatives of 106 emerging countries, including all petroleum exporters, unanimously agreed to expand the International Monetary Fund loan facilities, due to pressures being exerted by the high petroleum prices. The initial contribution by the Fund, in 1974, was on the order of $3.4 billion, and this figure may be expanded to from $10 billion to $15 billion in 1975.

Cooperation in Programs for Normalizing and Orienting International Trade in Raw Materials

The example and success achieved by OPEC has stimulated other raw-material-producing and agricultural countries to become united in more or less similar associations and organizations, which will enable them to normalize the prices of their export products and to establish quotas among them, thus correcting the international market's anarchical status, which has been traditionally taken advantage of by the great consumer nations. In this regard, OPEC members individually, and the organization proper, could press ahead programs oriented toward strengthening the associations of countries having the same objectives.

Venezuela's government has already taken a step in this direction by offering financing to coffee-producing countries, to enable them to stock their reserves and thus secure the corresponding prices. For the same purpose, representatives of Venezuela's government will possibly meet with the governors of Jamaica, Guyana, and Surinam, principal producers of bauxite. An association of this type already exists for copper producers.

We must bear in mind, however, that the development of an association such as OPEC, which involves only a small group of producers, is much easier than an association of producers of raw materials and agricultural products, whose members are more numerous and whose problems in attaining similar objectives are more complex. Notwithstanding, most Third World countries, exporters of raw materials and agricultural products, seem to be steadily oriented in this direction, and OPEC members can support them and collaborate in their projects.

Programs of Assistance to Third World Countries

Many Western world means of communication attempt to project the idea that the great volume of petrodollars flowing to the financial market can be absorbed only, in a large proportion and with due security, by industrialized nations. Simultaneously, intensive action is being undertaken by bankers and financiers of these nations aimed at securing recycling of the greatest possible amount of these funds toward these great powers.

Many financial and industrial companies of these nations offer greater security, due to their size, solvency, efficiency, and tradition, and OPEC countries need to obtain a return that at least compensates for the galloping world inflation. These are the main reasons for the recycling of a great portion of petrodollars toward Western industrial nations, especially in short-term investments in the Eurodollar market.

In this respect, in a report recently published in Brussels, the European Economic Council announced that investments abroad by OPEC countries, in the first ten months of 1974, have apparently reached a total of around $47 billion. Of this amount, $11 billion was invested in the United States, half in Treasury bills. Other placements of capital (equivalent to approximately $7 billion) were made in pounds sterling in the United Kingdom, principally in the form of deposits and government bonds. Around $16 billion to $17 billion was launched on the Eurodollar market, and $4 billion was invested in direct loans to England and Japan. Another $2 billion served for founding an "oil facility" at the International Monetary Fund; $2 billion was transferred to developing countries; and $4 billion was invested in the private sector, especially European.

The fact that nearly all these investments are over the short term seems to demonstrate the intention by Arab countries and other exporters, such as Venezuela, to gain time while they formulate more definite investment plans. In effect, investment plans of a more permanent nature, such as the industrial projects mentioned earlier, obviously require thoroughly detailed studies, and hence more time.

In their foreign investment programs, OPEC countries should contemplate recycling the bulk of their funds toward Third World countries, especially the most impoverished, for two obvious reasons, i.e., humanitarian and strategic.

As I have consistently reiterated throughout this book, petroleum-exporting countries have been forced, nearly overnight, to assume leadership in the great crusade for establishing a new, more balanced world economic system, and to break the bonds of dependency that subject Third World countries in general to the small group of industrial nations. To achieve this basic objective, OPEC countries should coordinate their maximum efforts—through direct investment and programs of economic aid to underdeveloped countries—thereby contributing to solving their great economic and social problems. This last measure, of a greatly humanitarian content, might be oriented through existing world organizations (such as the World Bank, the Agency for Reconstruction and Development, and the Inter-American Development Bank), through new organizations possibly created by OPEC, or else through OPEC directly.

For OPEC to be able to efficaciously collaborate in programs of such magnitude, which imply a large-scale expansion of its activities, it should take into account all the necessary administrative and managerial measures. In this sense, it pleased me to see that OPEC's meeting on January 24, 1975, was attended by Ministers of Foreign Relations and of Finance besides its traditional founding members (the Petroleum Ministers). This indicates that the idea is already taking root in OPEC, to the effect that its members must initiate wide-ranged, global, diplomatic, and strategic tactics, and therefore OPEC meetings must include attendance by high government officers of member countries, together with the petroleum experts, because OPEC must become, in view of world circumstances, an organization having much broader objectives and aims.

In turn, as a counterpart and in solidarity with the great common cause uniting them, nonpetroleum underdeveloped countries, beneficiaries of investment and economic assistance by OPEC countries, should provide all their support and collaboration to petroleum countries in their endeavor to maintain and consolidate petroleum prices and in achieving the other goals that I have outlined.

Investment Strategy

When referring to fundamental goals, I pointed to the prime importance of adequately investing the vast funds generated by petroleum in programs directed toward fully developing the economies and social and cultural levels of underdeveloped countries. As such programs require exhaustive study, not all these resources can be so invested immediately. Consequently, some must be temporarily placed abroad, taking very much into account diverse criteria.

I am not pessimistic over the maintenance of fair petroleum prices, because exporter countries have the elements available to defend them. However, we must bear in mind, I reiterate, that on the one hand petroleum is a nonrenewable resource, consequently limited in time, and on the other, great industrialized petroleum-consumer nations will exert themselves in developing alternative sources of energy.

Because of the foregoing, a general mobilization should develop in petroleum countries toward judiciously investing the highest possible proportion of their oil money locally, as it is not known how long this bonanza might last. A fundamental tool for adequately investing these funds is the elaboration of medium- and long-term national plans that should be vigorously carried out and rigorously supervised by the state, supported by leaders and the public in general.

Some petroleum-exporting countries already have experience in drawing up national plans and programs over the medium term. Venezuela, for example, has been carrying out five-year plans since 1959. However, these plans leave a lot to be desired insofar as their elaboration and execution is concerned, for reasons beyond the scope of these comments. Therefore, and in view of current circumstances, higher efficiency is essential in the preparation and execution of these plans.

Simultaneous to the preparation of the national plans, Venezuela's government (as well as governments of other petroleum-exporting countries) should make an extra effort to structure a foreign investment program—unprecedented in history—for absorbing surplus funds. This should be done along the following lines of criteria:

Protection of Investments

It seems obvious that Venezuela and other petroleum countries should attempt to avoid losses in their investments abroad. Because of the difficulties involved in control and supervision, pains must be taken in the preliminary study of all intended investment projects and in their supervision once they are put into practice. Special care should be taken to avoid "collaboration" by unscrupulous or incompetent foreign or national advisors, many of whom are becoming actively mobilized to take advantage of the situation as best as possible for themselves.

To a maximum extent countries should use national experts for carrying out the studies, analyses, and supervision of their investments. They should rely only when absolutely necessary on the assistance of foreign financial and technical advisors of known impartiality and moral and technical solvency.

By all possible means, countries must ensure that the selection of personnel for these projects, both local and foreign, is effected without interference from political groups. In this respect, they must endeavor to prevent the corruption to which vast investments of funds are vulnerable—especially abroad, where supervision and control thereof is more difficult. It must always be borne in mind that the funds to be invested belong to the state, and hence to the entire population.

Achievement of National Goals Supported and Abetted by Petroleum Funds

It seems obvious that the fundamental and ultimate use of vast petroleum funds should be in the expansion, welfare, and progress, at all levels, of the oil-exporting countries. In this regard, it would be ideal if a very high proportion (deducting the necessary investments and programs of assistance to other Third World countries) were invested locally to finance first-class projects for national expansion. As this should be the medium- or long-term goal, investments abroad should be classified into two large groups: temporary and permanent.

The first group, temporary investments, should be composed of funds that cannot be immediately invested locally and are not reserved for specific international policy plans. These funds should be placed over the short or medium term, in a substantially liquid form, to produce adequate return (at least to compensate for depreciation of the currencies in which it is invested), while awaiting return to its country of origin for final use in national, or eventually international, plans. Therefore, these resources are in fact a reserve for the more or less immediate future.

The second group, permanent foreign investments, although situated abroad, should directly or indirectly contribute to the development of national domestic plans, thereby contributing in some manner to strengthening the national economy at a cultural or social level, or at an international level of the investing country. An example is the development of regional or subregional plans of integration, which include the petroleum country contributing the investment and which, it is assumed, will be economically favored. In Venezuela an example is ALALC, the Subregional Andean Pact. Or investment might be made in various industries or organizations abroad that, due to their technology or degree of development in certain fields, might support domestic plans. We have an example of this last case in the acquisition by the Iranian government of 25 percent of Krupp Huettenwerke shares. This investment will enable Iran to rely on a further tool for developing substantial siderurgical projects within its country.

This type of orientation in investments abroad should help to belie alarmist assertions frequently appearing in the Western press, to the effect that OPEC countries will carry out an offensive aimed at absorbing a great number of industries in the developed world. This journalistic attitude is evident, for example, in *The Economist,* a reputable publication in which it has been asserted that the twelve OPEC member nations would, with their income, be able to buy all world industries within fifteen years; the New York Stock Exchange in nine years; all the gold in central banks in three years; all French and German industries in 324 days; IBM in 143 days; the Milan Stock Exchange in 50 days; Fiat in 3 days; and Olivetti in 15 hours.

The possible acquisition of certain companies in the industrial world by the governments of OPEC countries should not form part of an inverted process of domination or colonization, but rather should form part of carefully studied and calculated plans with the aim of complementing their national programs for industrial, agricultural, and housing development and so on. It therefore will not consist of an economic offensive similar to that exercised by United States multinational companies in Europe (and in the Third World) after World War II, which was dramatically described in J. J. Servan-Schreiber's best-seller *The American Challenge.*

Strengthening the Oil Industry Proper

One of the most important aims of all petroleum-exporting countries must be to fortify their own oil industry, which is the source of the prosperity and extraordinary bonanza they today enjoy. This is particularly necessary in Venezuela where, as we have seen especially in Chapter 8, the oil industry has been subject to a process of decline for more than a decade. Such a process, fundamentally evident in the decrease in reserves and in productive capacity, may be greatly detrimental to Venezuela in several aspects.

In the first place—apart from the circumstancial convenience of voluntarily decreasing production to avoid surplus that might negatively affect price levels, and to prevent waste through an excess of liquid revenue—a progressive decrease in production may cause the country to lose a considerable amount of income that it needs for achieving the great national goals mentioned earlier.

In the second place, it is logical to foresee that Western industrial powers will intensify their efforts to produce alternative sources of energy as soon as possible, which for reasons of space, I am unable to analyze. But I believe it pertinent to mention, as stated by the majority of experts on the subject, that it would prove to be extremely difficult to replace petroleum over a relatively short time (five or ten years), and in high proportion, by another source of energy. We must not forget that there are ecological factors that might restrict the intensive development of certain sources, such as nuclear energy, coal, and nonconventional petroleum (shale and schist). It therefore seems doubtful that by the end of this decade, and even by the mid-1980s, great industrial nations will have developed sufficient alternative sources of energy to become independent.

However, Venezuela and other OPEC countries should be fully conscious of the fact that, in the long run, their petroleum might be displaced by another energy source, and therefore, best advantage should be taken of it. Bearing in mind changing world factors and the present oil bonanza, they must plan the audacious and realistic action necessary for entering, within a relatively short time, into their phases of development.

For these reasons, Venezuela should prevent the systematic and continued decline of its productive capacity, as a consequence of the decrease in its reserves. To that effect, it should enter into a very ambitious program covering the inclusion of nonconventional oil reserves; it should elaborate ample plans for secondary recuperation at wells being presently exploited; and it should develop its large deposits of nonconventional heavy petroleum in the Orinoco Tar Strip.

Venezuela's action should be oriented not exclusively toward reactivating, strengthening, and developing its own industry, but rather toward becoming a *truly petroleum country,* i.e., developing a policy that will enable it to participate nationally and internationally in the industry's several fields, namely: production, refining, transportation, and marketing.

Diversification of Petroleum Countries as Producers of Alternative Sources of Energy

As mentioned previously, industrialized consumer nations have set production of alternative sources of energy as one of their main goals over the medium or long term, in order to avoid their dependency on petroleum. This goal is today a world objective, in view of nature's having distributed petroleum resources among only a small number of countries. The immense majority of countries either totally lack petroleum or have insufficient amounts to meet their own consumption.

In view of these limitations and of the high cost involved in developing new sources of energy, it seems logical that the small number of petroleum-exporting countries, which are accumulating a major part of world liquid resources, should join in the great effort being made to develop these alternative sources which, in any case, will be necessary, since present world petroleum reserves, and those that might be added in the future, will not—in the long run—be sufficient to adequately meet world needs.

In this manner, petroleum countries will not only be contributing to this constructive effort at a world level, but will be securing the possibility of continuing in the future as producers of energy, even when their own petroleum resources begin to decline. Special consideration must be given to the fact (frequently referred to by some spokesmen of exporting countries, especially the Shah of Iran and his Minister of Finance, Jahangir Amuzegar) that it is precisely the high petroleum prices that are fomenting the development and commercial exploitation of other sources of energy. If prices were to fall below present levels, other potential sources of energy, because of high exploitation costs, would not be able to compete with petroleum, and hence their development would suffer serious setbacks.

Venezuela should join this movement even though, like other exporting countries, it lacks the technological know-how to develop alone, within or outside of its own territory, other great sources of energy, such as nuclear or solar, which require refined and complex processes. But these petroleum countries do possess the necessary liquid funds, and by incorporating these funds with the technology of industrialized nations they could set the corresponding projects satisfactorily in motion. In this field, OPEC should play an important coordinating role, contributing its proved negotiating ability for dealing with the great consumer nations that possess the technology.

We must not overlook the fact that Venezuela possesses great sources of energy that require less advanced technological processes, such as hydro-electricity. It has important hydroelectric sources under process of development, such as the great Guri Dam, and therefore has accumulated a fine degree of experience in this field. Furthermore, it possesses vast potential resources of hydroelectric energy that would allow it to develop much greater projects within its own borders. Venezuela should also study and carry out important joint ventures with other Latin American countries, such as Brazil, which also has great potential resources of hydroelectric energy. It should endeavor to participate in the development of more complex sources of energy, at a world level, possibly supported by OPEC's coordinating capacity, mentioned previously.

Strengthening Export Countries' Status in World and Regional Organizations

I mentioned earlier that the world has lived through a clearly defined experience to date in this field: developed nations, principally because of their economic power, have taken a leading role in the administrative and decision-making bodies of several international organizations. This role is a consequence of their major financial contributions to these international organizations, in endowment, programs carried out, and fiduciary funds. Therefore, if petroleum-exporting countries are going to contribute important sums to existing international institutions, or to those that might be established in the future, for helping to improve the balance of payments and economic conditions of other countries, especially the most impoverished, they must demand a strong position in such organizations that will permit them, among other things, to supervise the administration of funds placed therein.

Venezuela, Helping to Strengthen Other Latin American Countries

Venezuela, within the general strategy of helping to strengthen the economy of all Third World countries, should first orient its action toward the Latin American countries, of which it forms a part. In this connection I would like to make the following comments:

Venezuela, by being an important petroleum-exporting country and because of its new financial power derived therefrom, is obliged to assume a leadership role in Latin America, whether it wishes to or not.

It therefore seems logical to state—without detriment to the fact that its petroleum policy is intimately coordinated with that of OPEC—that Venezuela is called upon to project a policy of financial cooperation toward Third World countries located in Latin America. The region's geographical vicinity and close historical and cultural ties lead it to being the natural area in which Venezuela should concentrate its policy of cooperation in forthcoming years.

A great part of the oil money received by petroleum-exporting countries, as we have seen, has been temporarily recycled toward developed nations (Europe and the United States) by means of short-term deposits and direct investments. Venezuela (as well as all OPEC members) must decide, over a relatively short term, how to invest surplus funds in a more permanent manner. Earlier in this chapter I referred to the criteria and general outlines which, in my opinion, should be pursued by petroleum-exporting countries in investing their excess funds abroad, which should be channeled toward Third World countries as far as possible. Within this policy, it seems logical that the recycling of petrodollars should be oriented toward countries and areas most closely associated to the individual OPEC countries. Therefore, it is only fair that Venezuela, in this general criteria, should orient its investments and foreign assistance toward Latin America.

Venezuela's government has started to actively direct its foreign policy toward expanding its relations with Latin American countries, and to that effect it has held meetings at presidential and technical levels with Central American, Caribbean, and South American countries. It has also programmed a meeting of all Latin American Presidents in Caracas in 1975.

These initiatives have led some press spokesmen to remark that Venezuela is aiming at becoming the leader of these countries. It is furthermore being said that this policy could create negative tensions in Latin America. Nothing could be more erroneous than these assertions, which reflect an ignorance of the genuine economic realities in Venezuela and in the region. Venezuela aspires exclusively to secure its economy, development, and petroleum achievements, and to cooperate internationally with all countries (especially with those of the region) in which it would be possible to develop activities of common interest, framed within the great cause of all Third World countries, which is to establish a new, more equalized, and fair international economic order.

This does not concern, then, an imposition of economic or political hegemony. Rather, it concerns a conciliation of economic interests. Therefore, Venezuela and Latin American countries are seeking a rapport in which to orient their policies toward sincere cooperation. An example is their united action in connection with the 1975 United States Trade Act and the economic cooperation that Venezuela has already started providing to Central American countries.

Venezuela does not wish to impose itself as leader in this region of the world or among underdeveloped countries in general. Nor does it wish to have any conflict with present international powers. Its current economic potential, however, and its desire to contribute to overcoming bonds of dependency oblige it to create a foreign policy with characteristics that emphasize aid to its less privileged neighbors in all of Latin America.

Summary

I have attempted in this book to reflect the image of what Venezuela's petroleum industry has been, what it is, and what it might become. At an international level, my concern has led me into discussing several subjects that make up the most delicate and disturbing aspects of present world economy and policy structures. I feel that however unconcerned people might be, they must at times reflect on matters concerning the dependency of underdeveloped countries on industrialized nations; the use of multinational corporations as a tool for economic domination; the alarming impoverishment of many Third World countries; the increasingly acute dependency of Western industrialized nations on raw materials; the current economic recession of these industrialized nations, etc.

These subjects are now under review by individuals concerned about Third World problems, to propitiate a new and more balanced international order. However, many political leaders of the developed world are nonetheless reluctant to accept the changes that economic realities and their social effects are dramatically pointing to as undeferrable. Even more serious is the fact that they react by expressing economic, political, and military threats, as if by such means they might reach in-depth solutions, which in reality are attainable only through peaceful understanding and comprehension among all nations.

It is quite possible that I have at times become carried away by the intensity with which these subjects are at present being debated, or that my viewpoints do not fully satisfy my readers' inquiring minds. However, when matters involved are so controversial, it is logical that not all opinions can be unanimous.

Besides narrating the mesmerizing history of Venezuela's petroleum industry, I have been motivated to point to events that are maintaining dangerous tensions in the world and to suggest some alternative measures that might, I feel, be adopted by petroleum-exporting countries in their capacity today as obligatory leaders of all Third World countries. The future of humanity will greatly depend upon the solutions given to these far-reaching problems.

Statistical Appendix

A word of caution: Historic statistical data in Venezuela are often unavailable, incomplete, or, sometimes, in conflict with those from other government sources. Although macroeconomic data have been published regularly from 1950 and work force data from 1960 by the Central Bank of Venezuela, in some cases it is difficult to make comparisons for more than three years.

POPULATION
(Thousands of persons)

Years	Population	Source
1498	400	Estimated by Alejandro Humboldt
1800	800	Estimated by Alejandro Humboldt
1810	802	Estimated by Alejandro Humboldt
1825	660	Official Census
1838	887	Official Census
1839	945	Estimated by Agustin Codazzi
1844	1,219	Official Census
1847	1,273	Official Census
1854	1,564	Official Census
1857	1,788	Official Census
1873	1,784	1st Official Census
1881	2,075	2nd Official Census
1891	2,324	3rd Official Census
1920	2,412	4th Official Census
1926	3,027	5th Official Census
1936	3,367	6th Official Census
1941	3,851	7th Official Census
1950	5,035	8th Official Census
1961	7,524	9th Official Census
1971	10,778	10th Official Census

Geometric growth rate 1950–1971 3.7%
1961–1971 3.7%

GROSS DOMESTIC PRODUCT BY MAJOR EXPENDITURE COMPONENTS, 1950–1973
(Millions of dollars at current market prices)

Years	Consumption			Gross fixed investment	Inventory change	Net foreign investment	GDP
	Private	Government	Total				
1950	1,472	381	1,853	656	28	−24	2,514
1951	1,568	414	1,982	752	30	4	2,769
1952	1,502	436	1,938	955	94	−7	2,983
1953	1,732	458	2,180	1,020	8	−30	3,178
1954	1,865	505	2,370	1,189	12	−54	3,517
1955	2,203	534	2,737	1,050	50	−30	3,806
1956	2,569	544	3,113	1,214	55	−113	4,269
1957	3,320	609	3,929	1,417	72	−514	4,904
1958	3,173	851	4,024	1,420	83	−123	5,354
1959	2,865	1,018	3,883	1,600	97	−63	5,518
1960	3,417	877	4,294	1,142	−68	744	6,112
1961	3,506	864	4,370	1,020	81	957	6,428
1962	3,827	845	4,672	1,104	135	1,119	7,030
1963	4,059	984	5,043	1,190	113	1,317	7,663
1964	4,649	1,007	5,656	1,493	310	1,026	8,485
1965	5,164	1,115	6,279	1,660	221	870	9,030
1966	5,412	1,218	6,630	1,769	109	900	9,408
1967	5,665	1,293	6,958	1,888	124	941	9,911
1968	5,535	1,350	6,885	2,438	355	700	10,378
1969	5,980	1,462	7,442	2,727	361	712	11,242
1970	6,655	1,640	8,295	2,696	736	655	12,382
1971	7,005	1,892	8,897	3,134	502	1,031	13,564
1972	7,813	2,078	9,891	3,723	484	1,021	15,119
1973	8,536	2,325	10,861	4,593	598	2,337	18,289

GROSS DOMESTIC PRODUCT BY MAJOR PRODUCTION SECTORS, 1950–1973
(Millions of dollars of constant prices)

Years	Petroleum	Agriculture	Manufacturing and oil refining	Services (public and private)	Others	Total
At 1957 market prices						
1950	904	241	303	1,363	219	3,030
1951	1,029	275	323	1,480	277	3,384
1952	1,094	295	376	1,552	313	3,630
1953	1,071	305	422	1,719	338	3,855
1954	1,153	306	481	1,890	396	4,226
1955	1,313	322	540	2,012	414	4,601
1956	1,500	344	591	2,158	494	5,087
1957	1,699	359	659	2,448	513	5,678
1958	1,597	375	708	2,531	542	5,753
1959	1,703	391	814	2,712	586	6,206
1960	1,744	473	797	2,851	591	6,456
1961	1,784	476	845	3,140	536	6,781
1962	1,956	501	921	3,486	536	7,400
1963	1,985	529	988	3,882	526	7,910
1964	2,080	571	1,111	4,310	606	8,678
1965	2,123	606	1,202	4,616	645	9,192
1966	2,060	632	1,214	4,810	687	9,403
1967	2,165	666	1,280	4,960	711	9,782
1968	2,209	699	1,353	5,220	821	10,302
1969	2,196	721	1,422	5,486	839	10,664
At 1968 market prices						
1968	2,131	740	1,242	5,124	1,141	10,378
1969	1,985	814	1,346	5,797	1,178	11,120
1970	2,188	844	1,481	6,202	1,248	11,963
1971	2,041	837	1,575	6,444	1,315	12,212
1972	1,897	854	1,728	6,844	1,445	12,768
1973	1,985	903	1,861	7,161	1,615	13,525

OIL CONTRIBUTION TO THE GROSS DOMESTIC PRODUCT, 1960–1973
(Millions of dollars at constant prices)

Years	Oil and gas contribution	Refining	Oil GDP	GDP	Percent to total
At 1957 market prices					
1960	1,744	99	1,843	6,344	29.1
1961	1,784	106	1,890	6,466	29.2
1962	1,956	117	2,073	6,842	30.3
1963	1,885	120	2,105	7,100	29.6
1964	2,000	122	2,201	7,697	28.6
1965	2,123	127	2,250	8,087	27.8
1966	2,060	129	2,189	8,254	26.5
1967	2,165	129	2,294	8,610	26.6
1968	2,209	134	2,343	9,106	25.7
At 1968 market prices					
1968	2,131	410	2,541	10,378	24.5
1969	1,985	417	2,402	11,120	21.6
1970	2,188	463	2,651	11,963	22.1
1971	2,041	441	2,482	12,212	20.3
1972	1,897	434	2,331	12,768	18.3
1973	1,985	480	2,465	13,525	18.2

VENEZUELA'S WORK FORCE, 1960–1972
(Thousands of persons)

Years	Oil industries	Agriculture, livestock, etc.	Non-oil industries	Commerce	Services	Others	Total employees	Work force	Unemployment rate (%)
		Working population by sectors							
1960	41	732	253	252	499	254	2,031	2,328	12.8
1961	37	730	271	239	540	235	2,052	2,393	14.2
1962	35	722	299	260	554	228	2,098	2,459	14.7
1963	34	713	325	286	570	234	2,162	2,528	14.5
1964	33	706	373	345	598	269	2,324	2,602	10.7
1965	32	701	409	400	613	289	2,444	2,681	8.8
1966	29	697	414	428	643	305	2,516	2,764	9.0
1967	27	694	444	455	667	321	2,608	2,852	8.5
1968	25	694	473	492	730	401	2,815	2,978	5.5
1969	24	706	496	520	734	418	2,898	3,104	6.6
1970	24	649	564	565	803	428	3,033	3,228	6.0
1971	23	656	574	585	807	475	3,120	3,312	5.8
1972	23	673	624	606	830	484	3,240	3,425	5.4

DOMESTIC LABOR INCOME, 1960-1972

	Total oil industry wages* (Million of dollars)	Domestic labor income	Average worker's income Oil ind. (Dollars per year)	National
1960	282	2,867	6,156	1,411
1961	255	2,961	6,634	1,443
1962	233	3,031	6,524	1,445
1963	235	3,344	6,950	1,546
1964	257	3,718	7,716	1,600
1965	263	3,953	8,271	1,618
1966	285	4,242	9,047	1,686
1967	271	4,563	10,000	1,760
1968	254	4,870	9,976	1,730
1969	246	5,150	10,018	1,777
1970	273	5,004	11,377	1,650
1971	275	5,524	16,607	1,770
1972	288	6,123	12,340	1,890

*Includes bonus, living allowance, etc.

OIL WELLS AND AVERAGE PRODUCTION, 1950-1973

Years	Total wells	Producing wells	Average production per well (barrels per day)
1950	10,756	6,934	216
1955	15,781	9,714	222
1960	21,827	9,933	287
1965	24,609	11,641	298
1966	24,968	11,416	295
1967	25,286	11,599	305
1968	25,709	13,338	270
1969	26,086	11,792	305
1970	26,729	12,128	306
1971	27,346	11,097	320
1972	27,821	11,299	285
1973	28,213	12,658	266

OIL: RESERVES AND PRODUCTION, 1917–1973
(Millions of barrels daily)

Years	Reserves	Production	Refined	Average life (years)
1917	. . .	0	0	
1922	. . .	0	0	
1925	. . .	0	0	
1930	. . .	0.3	0	
1935	. . .	0.4	0	
1940	. . .	0.5	0	
1945	7.0	0.5	0	22
1950	8.7	1.5	0.3	16
1955	12.4	2.2	0.5	16
1960	17.4	2.8	0.9	17
1965	17.2	3.5	1.2	14
1966	16.9	3.4	1.2	14
1967	16.0	3.5	1.1	12
1968	15.7	3.6	1.2	12
1969	14.9	3.6	1.2	11
1970	14.0	3.7	1.3	10
1971	13.8	3.5	1.2	11
1972	13.9	3.2	1.1	12
1973	14.1	3.4	1.3	11

NATURAL GAS: RESERVES AND PRODUCTION, 1950–1973
(Billion cubic feet)

Years	Reserves	Production	Flared	Reinjected	Average life (years)
1950	16.9	1.5	1.3	0.1	33
1955	23.3	2.4	1.7	0.4	32
1960	33.6	3.0	1.5	1.1	46
1965	30.0	3.9	1.5	1.7	37
1966	29.2	4.0	1.5	1.8	36
1967	27.8	4.4	1.7	1.9	30
1968	26.8	4.5	1.6	2.0	30
1969	32.9	4.6	1.7	2.0	36
1970	32.8	4.7	1.8	1.9	33
1971	38.4	4.6	1.6	2.0	41
1972	38.8	4.4	1.4	2.0	43
1973	37.9	4.8	1.4	2.1	39

OIL REFINING, 1917–1973
(Thousands of barrels daily)

Years	Oil	Gasoline	Fuel oil	Desulf. fuel oil	Diesel and gas oil	Others
1917	. . .					
1922	1					
1925	3		N O T A V A I L A B L E			
1930	14					
1935	25					
1940	73	6	52	. . .	13	1
1945	89	8	63	. . .	15	3
1950	250	40	141	. . .	48	21
1955	537	71	284	. . .	117	64
1960	882	106	512	. . .	156	109
1965	1,175	130	734	. . .	195	116
1966	1,174	139	714	. . .	194	128
1967	1,166	138	701	. . .	185	140
1968	1,186	157	707	. . .	183	137
1969	1,156	165	706	. . .	151	134
1970	1,292	179	815	. . .	151	145
1971	1,245	188	590	188	158	116
1972	1,125	183	541	137	148	119
1973	1,303	191	628	204	161	122

OIL AND REFINED PRODUCTS: EXPORTS FROM VENEZUELA, 1917–1973
(Thousands of barrels daily)

Years	Oil	Gasoline and naphtha	Fuel oil	Desulfur. fuel oil	Diesel and gas oil	Others	Total exports
1917	0						0
1922	5						5
1925	50		N O T A V A I L A B L E				50
1930	368						368
1935	380						380
1940	429						429
1945	870						870
1950	1,240	23	129	. . .	31	1	1,424
1955	1,619	29	260	. . .	92	24	2,024
1960	1,998	45	480	. . .	122	40	2,685
1965	2,332	60	669	. . .	124	68	3,253
1966	2,263	65	654	. . .	123	77	3,182
1967	2,429	62	656	. . .	123	91	3,361
1968	2,455	79	639	. . .	110	85	3,368
1969	2,476	74	691	. . .	89	81	3,411
1970	2,435	93	745	. . .	107	90	3,470
1971	2,314	103	545	168	93	59	3,282
1972	2,132	85	514	185	92	57	3,065
1973	2,124	87	609	207	70	53	3,150

VENEZUELA'S CRUDE OIL EXPORTS, 1917–1973
(Thousands of barrels daily)

Years	United States	Canada	Europe	Japan	Others	Netherlands Antilles	Total
1917							0
1922							5
1925			NOT AVAILABLE				50
1930							368
1935							380
1940							429
1945	203	16	6	...	12	561	798
1950	286	84	61	...	58	751	1,240
1955	393	185	142	...	143	756	1,619
1960	558	200	271	...	265	703	1,997
1961	501	224	366	0	221	726	2,038
1962	569	233	431	5	238	744	2,200
1963	588	242	414	6	266	726	2,242
1964	583	281	403	6	324	754	2,351
1965	559	232	424	8	349	760	2,332
1966	578	169	387	9	377	743	2,263
1967	547	248	465	7	400	762	2,429
1968	569	279	440	9	454	704	2,455
1969	537	299	450	11	457	722	2,476
1970	539	329	405	11	376	775	2,435
1971	584	337	367	9	336	681	2,314
1972	567	324	290	8	302	641	2,132
1973	680	300	239	10	309	586	2,124

OIL BY-PRODUCTS: EXPORTS FROM VENEZUELA AND NETHERLANDS ANTILLES, 1945–1973
(Thousands of barrels daily)

Years	United States	Canada	Europe	Japan	Others	Total
1945	42	3	91	Not avail.	417	553
1950	328	21	226	Not avail.	274	849
1955	441	41	253	1	402	1,138
1960	684	59	250	3	301	1,297
1961	738	54	264	5	297	1,358
1962	766	59	324	19	271	1,439
1963	750	63	398	20	232	1,463
1964	837	80	336	25	247	1,525
1965	925	120	291	34	220	1,590
1966	976	120	234	41	211	1,582
1967	990	143	211	60	206	1,610
1968	984	145	154	55	224	1,562
1969	1,121	143	122	26	220	1,632
1970	1,295	150	111	22	227	1,805
1971	1,212	91	110	0	213	1,626
1972	1,196	91	78	1	186	1,552
1973	1,379	76	98	0	181	1,734

Index

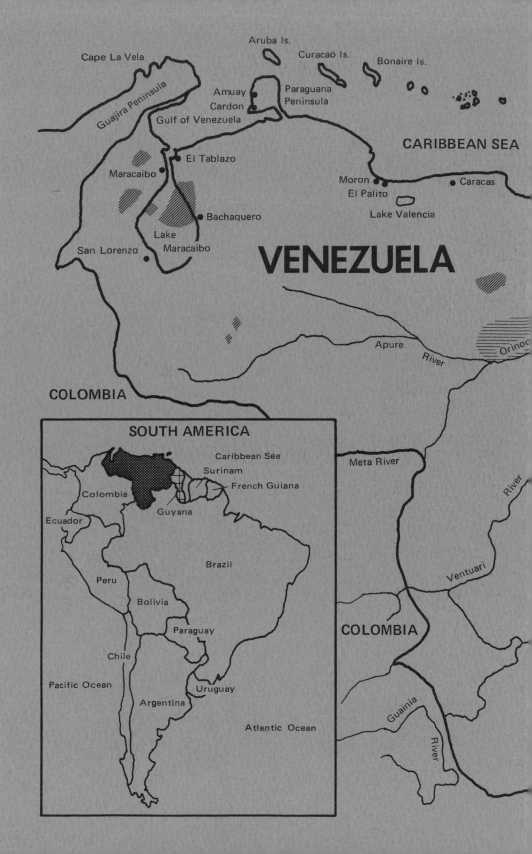